A Primer of Molecular Biology

CURRENT TOPICS IN CARDIOLOGY

Series Editor

Shahbudin H. Rahimtoola, MB, FRCP
George C. Griffith Professor of Cardiology
and Professor of Medicine
Chief, Division of Cardiology
University of Southern California
Los Angeles, California

Acute Myocardial Infarction
Bernard J. Gersh, MB, ChB, DPhil, FRCP
Shahbudin H. Rahimtoola, MB, FRCP

Cardiac Electrophysiology and Arrhythmias
Charles Fisch, MD
Borys Surawicz, MD

A Primer of Molecular Biology
Robert Roberts, MD
Jeffrey Towbin, MD
Thomas Parker, MD, PhD
Roger D. Bies, MD

A Primer of Molecular Biology

Robert Roberts, MD
Professor of Medicine and Cell Biology
Chief of Cardiology
Baylor College of Medicine
Houston, Texas

Jeffrey Towbin, MD
Assistant Professor of Pediatrics
Pediatric Cardiology and the Institute
for Molecular Genetics
Baylor College of Medicine
Houston, Texas

Thomas Parker, MD, PhD
Assistant Professor of Medicine
University of Toronto
Toronto, Ontario, Canada

Roger D. Bies, MD
Fellow, Section of Cardiology
Baylor College of Medicine
Houston, Texas

Elsevier
New York • Amsterdam • London • Tokyo

No responsibility is assumed by the Publisher for any injury and/or damage to persons or property as a matter of products liability, negligence or otherwise, or from any use or operation of any methods, products, instructions or ideas contained in the material herein. No suggested test or procedure should be carried out unless, in the reader's judgment, its risk is justified. Because of rapid advances in the medical sciences, we recommend that the independent verification of diagnoses and drug dosages should be made. Discussions, views, and recommendations as to medical procedures, choice of drugs, and drug dosages are the responsibility of the authors.

Elsevier Science Publishing Co., Inc.
655 Avenue of the Americas, New York, New York 10010

Sole distributors outside the United States and Canada:
Elsevier Science Publishers B.V.
P.O. Box 211, 1000 AE Amsterdam, The Netherlands

This book is printed on acid-free paper.

Library of Congress Cataloging-in-Publication Data

A primer of molecular biology/Robert Roberts . . . [et al.].
p. cm.—(Current topics in cardiology)
Includes index.
ISBN 0-444-01657-0 (hard cover : alk. paper)
I. Heart—Molecular aspects. I. Roberts, Robert, 1940–
II. Series.
[DNLM: 1. Cardiovascular Diseases. 2. DNA, Recombinant.
3. Heart—physiology. 4. Molecular Biology. WG 100 P953]
QP114.M65P75 1992
612.1′7—dc20
DNLM/DLC 92-14601
for Library of Congress CIP

Current Printing (last digit):
10 9 8 7 6 5 4 3 2 1

Manufactured in the United States of America

Contents

Foreword to the Series

Knowledge about molecular biology has progressed at a rapid pace over a short period of time. Advances in this field are yielding new developments in the diagnosis and treatment of cardiovascular diseases.

One goal of our CURRENT TOPICS IN CARDIOLOGY series is to present comprehensive, in-depth updates of information in selected areas of cardiology. In the case of molecular cardiology—a new field—there has been a need for a "primer." Dr. Roberts and his colleagues have done an excellent job of providing this kind of brief introductory book for "students." Consistent with other volumes in this series, it is hoped that this work will prove valuable to practicing physicians and specialists, investigators, and academicians.

Shahbudin H. Rahimtoola, MB, FRCP

Preface

In a recent plenary address to the American College of Cardiology, Dr. Leroy Hood stated that he expected the application of the techniques of recombinant DNA and molecular biology to medicine would advance the field more in the next twenty years than all of the progress that has been made in the past 2,000 years. Cardiology, as a discipline, has been one of the last to embrace the techniques of molecular biology, yet the advances made in the past five years have indeed been dramatic. Thrombolytic therapy, introduced with a bang through the use of recombinant tissue plasminogen activator (rt-PA), is now part of the routine armamentarium in the treatment of patients with acute myocardial infarction. Tissue plasminogen activator was the first drug made with the use of recombinant DNA techniques to be used in cardiology and was approved by the Food and Drug Administration (FDA) in 1989. In the short time since then, five other drugs made by these techniques have been added to the treatment of patients with ischemic heart disease—hirudin, superoxide dismutase, urokinase, prourokinase, and multiple mutant forms of rt-PA. The FDA has approved fewer than 100 drugs that are made by recombinant DNA techniques, but estimates that by the year 2000 over 50% of all drugs will be made by these techniques. The ability to genetically engineer designer drugs, apart from being fashionable, is likely to dictate future trends in both the prevention and treatment of heart disease. The heart was not the organ of envy for the molecular biologist any more than it was for the physician scientist who wanted to apply the techniques of molecular biology, in large part because its cells (myocytes) do not possess the ability to proliferate. The cardiologist did not have the "carrot of enticement" as did the oncologist with his uncontrolled proliferative neoplasms, but even that has changed. This year alone, cardiologists will perform coronary angioplasty on 350,000 individuals, of whom 40% will return within six months because of an uncontrolled proliferative smooth muscle lesion referred to by us as restenosis.

In the short time of utilizing these techniques, we have also made considerable progress with respect to the heart itself. We have shown that despite the heart not being able to proliferate, it does indeed participate in the growth response not only of increased protein synthesis and muscle mass (hypertrophy), but is also capable of changing its phenotype. Several genes responsible for proteins, such as myosin and actin, participate in this growth response and express genes that give rise to fetal proteins that have not been expressed since in utero. This gives the heart the exciting potential (and the cardiologist the renewed hope) that myocytes, since they can express certain genes that they have not hitherto utilized throughout their lifetime, may also be able to recall the genes that once induced myocyte proliferation in utero. After all, cardiac cells have the same genes as all of the other cells of the body and were capable of proliferation in utero; therefore, inherently they have the ability to proliferate just like those of the liver and the skin. Would it not be exciting to have the ability to change the scar that forms following infarction to that of the contractile, fleshy muscle with the use of implanted cardiac myocytes or promoter-driven gene expression converting the undifferentiated fibroblast of the initial healing phase to that of contractile myocytes? The potential for more specific etiological diagnosis and cure of primary cardiomyopathies, particularly those due to inherited diseases such as hypertrophic cardiomyopathy and idiopathic dilated cardiomyopathy, is obvious. Isolation of the gene, elucidation of the pathophysiology of the disease, and development of more appropriate therapy including gene transfer and gene replacement beckons the cardiologist to make haste in applying the techniques of molecular genetics. Understanding the molecular basis for these primary defects is likely to provide important clues fundamental to cardiac growth and hypertrophy. This was best summarized by William Harvey in 1657 when he said "Nature is nowhere accustomed more openly to display her secret mysteries than in cases where she shows traces of her workings apart from the beaten path; nor is there any better way to advance the proper practice of medicine than to give our minds to the discovery of the usual law of Nature by careful investigation of cases of rare forms of diseases. For it has been found that in almost all things, that what they contain of useful or applicable nature is hardly perceived unless we are deprived of them, or they become deranged in some way."

The purpose of this primer is to provide a book in which the fundamentals and essential nomenclature would be appreciated by the complete novice. The information is provided such that (1) no prior background is required of the techniques of recombinant DNA and molecular biology, and (2) the book can be appreciated and understood by the trainee, clinical cardiologist, and investigator. It is an attempt to reach the clinical cardiologist and provide him with the necessary understanding to appreciate the innovative, diagnostic, and therapeutic trends that will evolve from the application of these new techniques to cardiovascular disorders. It is already evident to the practicing physician that when glancing through such time-honored clinical journals as *The New England*

Journal of Medicine or *The Annals of Internal Medicine,* one is somewhat at a loss if terms such as a Southern or Northern blot or a dot blot are not part of the reading vocabulary. Terms such as genomic DNA, complementary DNA (cDNA), and gene expression are now used routinely in clinical journals. The first chapter, *The Fundamental Discoveries and Unique Features of Recombinant DNA Techniques,* is an historical account highlighting four important discoveries that were pivotal to the development of recombinant DNA techniques. This is paraphrased from a lecture that one of the authors (Roberts) has given to varied audiences, to which the response has been that it better prepared them for understanding the fundamental building blocks required for these new techniques. Chapter 2, *Essentials of Nucleic Acids and Proteins,* is a brief presentation of the essentials of nucleic acid in preparation for the subsequent chapter (*Techniques of Molecular Biology*), which describes the more commonly utilized techniques of recombinant DNA and molecular biology. It includes techniques such as Southern and Northern blotting; isolation of RNA and DNA; and the techniques of cloning, sequencing, and polymerase chain reaction (PCR). These techniques are not provided as a protocol with the intent of teaching the reader to perform them, but rather to provide the rationale, purpose, and limitation of the technique.

The major response of the heart and practically the only long-term response to most forms of injury is that of cardiac growth and hypertrophy. This has been a very active area for research and is likely to remain so for the next decade. Therefore, we thought it important to present some of the experimental data fundamental to future implications that are likely to arise and be of diagnostic or therapeutic value in the future. Recognition that a variety of growth factors can indeed induce cardiac growth and alter its phenotype is likely to be common knowledge in clinical circles in the immediate future. It is highly likely that the 1990s will be remembered as the decade during which the molecular basis for cardiac growth and hypertrophy was elucidated. There is little doubt that this will be the decade in which many inherited diseases of the heart, particularly primary myopathies such as hypertrophic cardiomyopathy and myotonic dystrophy, will be elucidated and the mutation and molecular defect defined, if not corrected. The rapid pace of advancement of molecular genetics is exemplified by the most recent advance; namely, the isolation of the gene responsible for hypertrophic cardiomyopathy and, even more recently, for myotonic dystrophy. The power of positional cloning (reverse genetics) together with the recent addition of techniques such as PCR, chemical cleavage, and gradient denaturing gel electrophoresis in combination with the human genome project, will markedly advance the application of techniques of molecular genetics. Most cardiologists have considerable knowledge of the sarcomere and its contractile proteins, but are less likely to have a working knowledge of the cytoskeletal proteins that scaffold the sarcomere and provide the overall architecture of the myocyte. Several of these proteins, such as ankyrin, α-actinin, titin, tubulin, filamin, dystrophin, vinculin, and spectrin, many of which have now been cloned and sequenced, are known to play a very important role in healing and remodeling following cardiac injury such as myocardial infarction.

A brief description of the nature of these proteins and their corresponding gene and function is covered in Chapter 5, *Molecular Biology of Contractile and Cytoskeletal Proteins.* A family of proteins that is likely to play a major role in the immediate and long-term future contains those referred to as cardiac ion channels. At the present time, sudden cardiac death takes one person per minute, a total of 450,000 per year, due primarily to cardiac arrhythmias. Antiarrhythmic drugs, most of which are directed to interrupt cardiac sodium influx, have been most disappointing and recently shown to be deleterious. The application of the techniques of recombinant DNA and molecular biology provide us the ability to isolate, clone, and characterize the structure-function relationship of the sodium, potassium, and calcium channels of the heart, which should lead to more appropriate genetically engineered drugs that are more effective and have fewer side effects. It is hoped that since cardiac channels have sequences specific to the heart, it will be possible to develop therapies specific for cardiac channels that will not have the deleterious side effects due to binding to channels in organs such as the brain and skeletal muscle.

It is our hope that this primer will provide the student, house officer, practicing physician, or investigator with the armamentarium to appreciate and participate in the exciting and rapid growth that will evolve over the next decade from the application of these new techniques to cardiovascular disorders.

Robert Roberts, MD
Jeffrey Towbin, MD
Thomas Parker, MD, PhD
Roger D. Bies, MD

Acknowledgments

We would like to extend our appreciation to Dr. Antonio Gotto (Chief of Medicine), the faculty, and all of the Fellows who have come through our training program. We are particularly grateful to the American Heart Association for their support of the Bugher Foundation Center for Molecular Biology, and to the NHLBI for their support of the SCOR in Molecular Biology of Heart Failure.

The Bugher Foundation Center Training Program provided a unique opportunity whereby the Fellows in Cardiology and the faculty were brought together to train and learn the techniques of molecular biology. It was uniquely designed to train physician/scientists in these techniques, while at the same time bringing together investigators with expertise in basic and clinical research. The latter provided a forum to update and educate junior and senior faculty and served as a stimulant to apply these new techniques to clinical research.

We are particularly grateful to Ben Perryman and Michael Schneider in the molecular cardiology unit, and the senior clinical faculty of Craig Pratt, Miguel Quinones, Mario Verani, James Young, and Albert Raizner, who made the experiment a success. We now have several junior clinical investigators—Neal Kleiman, Douglas Mann, Roberto Bolli, and William Zoghbi—who, too, believe that the integration of these techniques with clinical expertise will facilitate future contributions. Most of our work was made easy by the administrative secretaries, Debora Weaver and Alex Pinckard, who through their diligence, energy, and patience made it possible for this to go to print. In conclusion, we would also like to thank certain individuals who were not only supportive, but often went without because of our work; namely, our wives and children.

Chapter 1

The Fundamental Discoveries and Unique Features of Recombinant DNA Techniques

The widespread application of the techniques of molecular biology and recombinant DNA throughout medicine has challenged the clinician to respond to the beckoning call of an appreciation of the terminology and the potential impact of these techniques. While these techniques are firmly entrenched in many specialties of medicine, such as genetics and endocrinology, cardiology has yet to embrace them fully. Several factors have probably played a role in the cardiologist's lack of utilization of the techniques of recombinant DNA: (1) the cardiologist is heavily burdened with clinical service; (2) the nature of the subspecialty is such that it requires extensive training in procedures that are labor intense, such as catheterization; and (3) the techniques of molecular biology are not a natural extension of those techniques routinely used in cardiology, as they might be for a subspecialty such as clinical genetics or oncology. During the years in which recombinant DNA was developed, the molecular biologist did not become interested in the heart for several reasons that were understandable at the time and to some extent remain so today: (1) cardiac myocytes are terminally differentiated and do not proliferate, and thus are not the first love for the molecular biologist; (2) the heart is such a vital organ that most mutations tend to be lethal, and thus genetic cardiac disorders have not provided information to the molecular biologist as have mutations in other organs; (3) the heart seldom develops neoplasms, an uncontrolled proliferative form of growth from which one may learn relevant clues to cardiac growth; and (4) until recently cardiac tissue could not be obtained readily and rapidly enough to provide intact RNA and DNA.

The large number of inborn errors of metabolism have of necessity as well as of interest brought the endocrinologist and the molecular biologist together, and for the same reason there has been a mutual collaboration between the oncologist and the molecular biologist. The slow embracement of the techniques of molecular biology as applied to the heart by both the molecular biologist and the cardiologist has led to a lack of knowledge on the part of faculty and trainees, to say nothing about the scarcity of M.D. investigators trained in molecular cardiology. Recognition of this deficit was the main reason the American Heart Association in 1985 funded the three Bugher Foundation Centers for Molecular Biology in the Cardiovascular System (1) in the United States to train cardiac fellows in molecular biology. In 1991 three more were funded, and in this short time of 6 years, several Divisions of Cardiology have become significantly involved in molecular biology of the heart and are embracing the techniques with great expectations for the future. Progress has been rapid in applying these techniques to the heart, and the results are very encouraging, as is the impact. Interest on the part of the clinician is accelerating at a very rapid pace. The first cardiac drug made by recombinant DNA techniques, appearing in 1983, was tissue-type plasminogen activator, (t-PA) which has been part of a major revolution in the therapy of acute myocardial infarction. In the short time since, several more cardiac drugs have appeared that are made by recombinant DNA, including hirudin, superoxide dismutase, urokinase, and pro-urokinase, and others are just around the corner.

In trying to acquaint the internist and cardiologist with the new techniques and terminology of molecular biology, the following overview, given as a lecture by one of the authors, has been well received and for that reason has been included with only slight modification (2).

Historical Perspective

The development of the techniques that led to modern molecular biology occurred in the 1970s and is synonymous with the birth of recombinant DNA technology. Despite the fundamental discoveries in the 1950s and 1960s, these techniques did not emerge until the late 1970s and early 1980s (3). Miescher isolated DNA for the first time in 1869, and in 1944 Avery et al. provided evidence beyond a doubt that DNA, rather than protein, is responsible for transferring genetic information during bacterial transformation (4). In 1953, Watson and Crick (5,6) proposed the double helix model for DNA structure, which was based on x-ray diffraction results by Franklin and Wilkins (7,8). It showed for the first time that the DNA molecule is double stranded, consisting of two complementary strands that are twisted around each other in a double right-handed α-helix. It was also shown that the four nucleotides (thymine, cytosine, adenine, and guanine) bind only to their complementary bases on the other strand. Thus, if the strands are separated and a new DNA molecule is synthesized and directed by either of the existing strands as a template, the new molecule will be identical to the original. The implication was obvious as a mechanism for passing on genetic information to subsequent generations.

This was followed in 1957 by the discovery by Kornberg (9) of DNA polymerase, an enzyme that is necessary for the synthesis of new DNA and one that is now used routinely in making DNA probes. Marmor and Doty and colleagues (10,11) showed that the double helix of DNA could be separated into individual strands (denatured), and reannealed or rehybridized. This observation, that the strands could separate and recombine (reanneal or hybridize) in an identical fashion, is very important. It is the basis of the specificity of most recombinant DNA techniques and is exploited in almost all of them. The strands, once separated, can again pair (hybridize) with each other or with other strands of DNA providing the bases are complementary, since adenine can only bind to thymine and cytosine to guanine. This process is used today to make synthetic DNA, referred to as synthetic oligonucleotides. The base pairing is also responsible for the specificity of DNA procedures, whereby if a piece of native DNA is labeled with an easily identifiable radioisotope or fluorescent compound (referred to as a DNA probe), it will only bind to a fragment of DNA if the sequences of bases are complementary. These fundamental observations, exploited in techniques such as Northern and Southern blotting (described later), are responsible for the specificity of these techniques. Nirenberg, Khorana, and their colleagues (12,13) documented the hypothesis of Crick that the genetic code was written in triplets of three base pairs coding for each amino acid. They subsequently discovered which triplet coded for each amino acid. The alphabet of DNA, made from just the four nucleotides, was now deciphered and the mechanisms for translation into the alphabet of protein—the amino acids—had been elucidated. Olivera et al. (14) discovered DNA ligase, the enzyme used to join DNA fragments together. Despite this wealth of fundamental knowledge that was absolutely essential for the development of recombinant DNA, the technique did not come forth in the 1960s and the field of molecular biology became less attractive, particularly to investigators interested in working with complicated mammalian systems.

There were several reasons accounting for this disinterest of the investigator in molecular biology, predominant among which were the large size of the DNA molecule and its monotonous nature, which made it difficult to detect a specific gene. The DNA molecule is a repeating unit of four bases, adenosine (A), cytosine (C), guanine (G), and thymine (T), and is the largest molecule of any organism. The human genome is known to contain about 3 billion base pairs that contain information that would more than fill a 500,000-page textbook. There is enough DNA to form about 10 million genes; however, it is estimated that only about 50,000 genes are required to code for a human being. Thus, less than 1% of our DNA is used to code for protein. There are 46 chromosomes, and each chromosome is a long, continuous DNA molecule. The chromosomes vary in size, but even chromosome 21, the smallest of them, contains more than 50 million base pairs. The difficulty facing the investigator in

Table 1.1 Discoveries Seminal to Modern Molecular Biology

1970	Discovery of specific restriction endonuclease
1970	Discovery of reverse transcriptase
1972/73	Development of the cloning technique
1975/77	Sequencing of nucleic acids

molecular biology in the 1960s was how to cut the DNA molecule into identifiable smaller pieces and how one could specifically select a piece of DNA of interest. Prior to 1977, sequencing of nucleic acids could be approached only indirectly through the protein. As a result of discoveries in the 1970s, the most difficult macromolecule of the cell to analyze, DNA, became among the most coveted and easiest to identify and to analyze. One can excise specific regions of DNA, obtain them in unlimited quantities, and determine the sequence of the nucleotides at a rate of more than 100 nucleotides a day. Specific genes can now be identified, genetically altered, and transferred back into cultured cells or into the germ line of animal and expression of the protein specifically determined.

The previously mentioned technical breakthroughs were, in large part, due to four seminal contributions (2) (Table 1.1). The first was the discovery and application of specific restriction endonucleases in 1970. They are to the molecular biologist what the scalpel is to the surgeon. These are enzymes (nucleases) that cut double-stranded DNA within the molecule (hence, endonucleases) rather than the ends at sites that are specific for each enzyme. The recognition sites for most enzymes are four to eight base pairs in length, with a few having recognition sites of only three base pairs and even fewer recognizing eight base pairs. Those restriction nucleases that have only three base pairs as their recognition site tend to cut the DNA into too many pieces, whereas those that recognize eight base pair sites cut the DNA into too few pieces. The restriction endonucleases are isolated from bacteria, where their normal function is to digest foreign DNA, restricting it from being incorporated into the genome; hence they are referred to as restriction endonucleases. There are well over 100 different types and the list is growing rapidly. It is thus possible to cut the DNA into fragments of a desired and consistent size, knowing specifically where each cut is performed. The ability to cut DNA into specific fragments is absolutely essential to all of the recombinant techniques, and more specifically was essential for the development of cloning. The existence of a DNA restriction endonuclease was first discovered by Arbor in 1962 (15); however, it was not until the work of Smith and Kelly (16,17) in 1970 that specific endonucleases were isolated and applied.

The second technical breakthrough was the independent discovery of reverse transcription in 1970 by two laboratories (18,19), which made it possible to generate from messenger RNA (mRNA) complementary DNA (cDNA). Next, (20), cloning was born, and the first recombinant molecule was made at Stanford. In 1971 (21) the first foreign DNA fragment was inserted into a plasmid to create chimeric plasmids, and it was shown that the plasmid could serve as a vector for inserting foreign DNA into an appropriate host, such as a bacterium. It was now possible to isolate mRNA, which codes for a specific protein, and, using reverse trancriptase, transcribe it into a single-stranded cDNA that could then be converted to double-stranded DNA, ligated to the DNA of a vector, and cloned in large quantities in an appropriate host. The fourth technical breakthrough occurred in 1977, when Sanger and Barrell (22,23) at Cambridge and Maxam and Gilbert (24) at Harvard independently developed rapid nucleic acid sequencing techniques. All of the important pieces were now in place to give birth to modern molecular biology and the widespread application of the techniques of recombinant DNA. It was not difficult to imagine how genetic engineering would provide a variety of molecules for both research and therapeutic purposes.

Unique Features of Recombinant DNA Technology

The techniques of recombinant DNA and molecular biology have ushered in a new era in research. Five major areas in which the techniques offer unique advantages (Table 1.2) over the techniques of existing disciplines are as follows: (1) the ability to perform in vivo structure-function analysis of a selected molecule or a portion thereof in the intact living cell or organism, (2) the molecular basis for the regulation of cardiac growth, (3) molecular genetics, (4) diagnostic in situ hybridization, and (5) the ability to generate in trace amounts large quantities of a protein present in the body that would not otherwise be available for therapeutic purposes, as well as the opportunity to genetically engineer drugs designed for maximal benefit with the least side effects.

Table 1.2 Unique Features of Recombinant DNA Techniques

The ability to perform in vivo structure-function analysis
The ability to unravel the molecular basis of cardiac growth
The ability to perform molecular genetics
The ability to perform diagnostic in situ hybridization
The ability to generate and genetically engineer specific therapies

In Vivo Structure-Function Analysis and Development of Specific Drugs

Prior to the development of recombinant DNA techniques, determination of the function of a specific molecule was often of necessity indirect, based on the results of in vitro experiments. The usual approach would be to isolate the specific protein of interest from the tissue with a high grade of purity and assess its kinetics with respect to substrate and product under controlled conditions. An example was determining the role of mitochondrial creatine kinase (CK) in energy transfer from the mitochondrion to the cytoplasm. Mitochondrial CK catalyzes the transfer of high-energy phosphate from adenosine diphosphate (ADP) to adenosine triphosphate (ATP) or creatine phosphokinase (CP), which are postulated to be important in the energy shuttle of ATP from the mitochondrion to the cytosol and myofibrils. Results of in vitro studies (25,26) showed that mitochondrial CK preferentially uses ATP as substrate, converting it to CP, which, being more soluble than ATP, diffuses throughout the cytosol to the myofibrils, where CK-MM, which prefers the substrate CP, converts it to ATP for immediate utilization. Despite multiple investigations over the past decades the results remain conflicting. A major question remains as to whether mitochondrial CK, located on the outer aspect of the inner membrane of the mitochondrion, does indeed facilitate transport of ATP. All attempts to answer this question to date have involved in vivo studies on isolated mitochondria or the sequelae of nonspecific inhibition in the isolated heart preparation. The difficulty with the latter in vivo approach is inability to selectively inhibit mitochondrial CK function without affecting other enzymes or molecules. In contrast, with recombinant DNA techniques, now that the mitochondrial CK gene has been cloned and sequenced, it is possible to modify the gene such that it codes for a nonfunctioning protein or, with a promoter, to overexpress mitochondrial CK in a living cell in which this is the only perturbation and observe the in vivo effects. In fact, such experiments with other molecules have now been performed in an intact animal, the transgenic mouse.

Other examples of studies being performed for in vivo structure-function analysis are those with the channel proteins responsible for the flux of sodium, calcium, and potassium ions. The mRNA for the selected channel is isolated, cloned, and injected into the oocyte and the ion flux monitored by patch clamping techniques. The opportunity to monitor precisely the function of a specific molecule in vivo has hitherto not been possible. The genetic engineering of a specific molecule, referred to as site-specific mutagenesis, can be used to clarify the pathophysiology of disease. The extent to which genetically engineered drugs will affect cardiac therapeutics of the future is so great that it may well be beyond our imagination.

Myocardial infarction, the number one killer in the western world, is due to a thrombus superimposed on an atheromatous plaque in the coronary artery. Myocardial damage as a result of coronary obstruction is the major determinant of morbidity and mortality and evolves over 4 to 6 hours from onset of symptoms. The recent intervention of thrombolysis has introduced a therapeutic revolution that has reduced hospital mortality by 25% to 50%. Initial studies were performed with streptokinase, followed by the introduction of recombinant-made tissue-type plasminogen activator (rt-PA). A cDNA containing all of the coding regions of the gene was expressed in bacterial and mammalian cell culture systems with and without portions of the cDNA removed, and the expressed product assessed for known functions, specifically lytic activity, fibrin affinity, and fibrin-dependent enhanced lytic activity (27) (Fig. 1.1A). Five domains were recognized to have specific functions that are coded by separate and autonomous portions (exons) of the gene: the finger domain and EGF domains are responsible for fibrin binding, $kringle_1$ and $kringle_2$ are responsible for enhancing lytic activity, and the light chain is the catalytic component of the enzyme and also contains the site for the binding of the plasmin inhibitor. Similar structure-function analyses have been performed on urokinase and single chain urokinase plasminogen activator (scu-PA), and by splicing together various portions of the genes, a variety of chimeric molecules have been generated. This paved the way for the ongoing research activities now in progress in several laboratories to make chimeric molecules that are therapeutically advantageous over the parent compound.

The hybrid thrombolytic agent developed by Haber probably represents the fourth generation of such agents (28) (Fig. 1.1B). It is the product of a fusion gene derived from three genes: one codes for the Fab fragment of an antibody to fibrin, another for the Fab fragment of an antibody to the t-PA inhibitor PA1, and the third for the catalytic unit of rt-PA. Preliminary studies show this molecule to be severalfold more potent in lysing clots and to have a higher affinity for fibrin. A most recent modification has been to remove from the rt-PA gene that portion coding for the finger EGF domains and the first kringle, which resulted in a molecule that has a half-life of 60 minutes and can be given as a single bolus injection, as opposed to the parent rt-PA, which has a half-life of 5 minutes and must be infused over 3 to 6 hours (29). The power of recombinant techniques to modify and specifically determine in vivo function of a specific molecule or portions thereof at a specific site provides the means to more rapidly develop specific drugs and to minimize or alter their side effects. It is estimated that well over 40 mutant forms of t-PA have now been developed.

Sudden death occurs in 30% to 40% of patients with myocardial infarction prior to reaching the hospital. Most of our drugs for ventricular arrhythmias and sudden death are directed against the sodium channel, which is responsible for sodium ion flux, and have serious side effects.

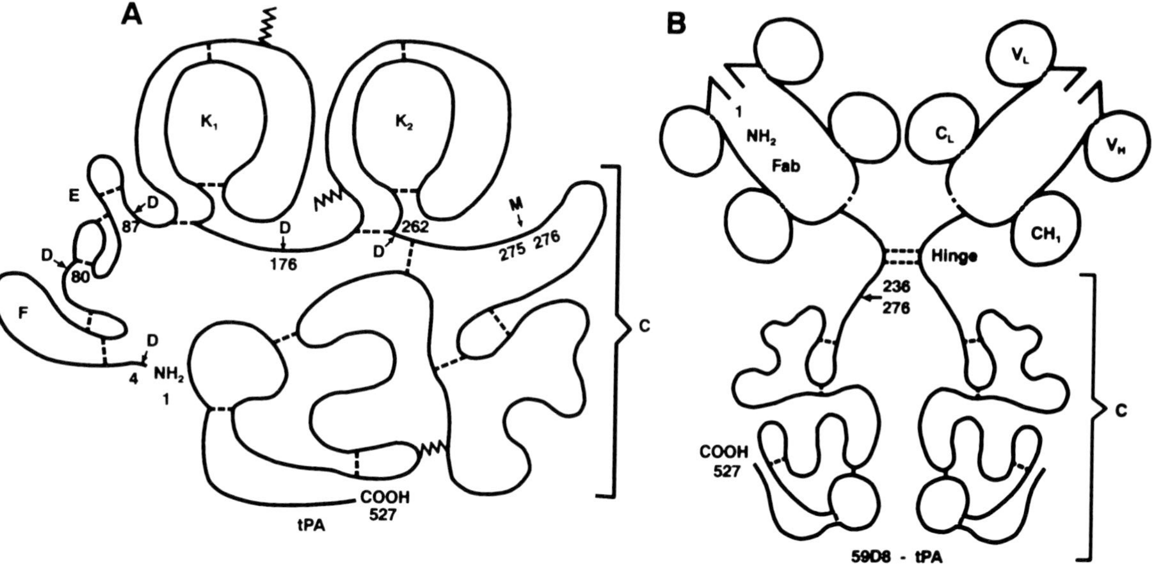

Figure 1.1 Two-dimensional representation of the structure of t-PA (A) and a chimeric plasminogen activator (B). (A) The α chain of r-PA is NH_2-terminal and the β chain is COOH-terminal to the plasmin cleavage site, P1, between residues 275 and 276. Functional domains are designated F, finger; E, epidermal growth factor–like; K, kringle, and C, catalytic. D indicates the limit of the individual functional domains deleted by Gething et al. (27). The dashed lines represent intrachain disulfide bonds, the zigzags *N*-linked oligosaccharides. (B) The Fab region of antifibrin antibody 59D8 in contiguous linkage to the β chain of r-PA between residues 236 of the 59D8 heavy chain (Gly in the construct) and Ile 276 of r-PA. VL and VH and CL and CH_1 denote variable and constant regions of the light and heavy chains. The molecule exists as a dimer linked at the immunoglobulin hinge region by a disulfide bond. *Reproduced with permission from Haber et al. Science 1989; 242:51–56. Copyright 1989 by the AAAS.*

The sodium channel is of diverse forms, with the cardiac sodium channel being different from that present in skeletal muscle or brain. The human cardiac sodium channel will soon be cloned and sequenced, and clearly drugs specifically engineered to influence selectively the cardiac sodium channel are likely to be more effective and associated with fewer side effects than the present drugs, designed without knowledge of the sodium channel protein. While rt-PA was the first drug made by recombinant techniques to be used in cardiology, in a period of just over 2 years four others have already been added: recombinant urokinase, prourokinase, superoxide dismutase, and hirudin. This is likely to be just the initial phase of the beginning of a revolution in recombinantly engineered therapy. The Food and Drug Administration, at the present time, has approved only about 60 drugs made by recombinant DNA, but estimates that by the year 2000, well over 50% of its approved drugs will be derived from recombinant techniques.

In Situ Hybridization

Diagnosis of disorders such as myocarditis continue to be made on clinical grounds, and seldom is it possible to isolate a virus by conventional techniques. A fundamental feature of single-stranded DNA or RNA is that it will only hybridize to its complementary DNA or RNA and the genome of each organism or individual is unique, thus, an RNA probe for a specific virus, many of which are now available, can be utilized to screen for viral RNA in human myocardial biopsies. Myocardial biopsies are a routine procedure in patients suspected of having cardiomyopathy but in whom the diagnosis cannot be determined by conventional means. Results of studies are already showing the diagnostic power of this technique in making the diagnosis of myocarditis caused by a virus (30,31). The recent development of the polymerase chain reaction (PCR) to amplify DNA or RNA to limits of several million copies or more, which can be detected by conventional techniques, provides a new horizon in the diagnosis of myocarditis and other forms of injury, such as immunological rejection. This, in essence, means that only one or two copies of viral RNA in a cell, which cannot be detected by conventional techniques, can now be amplified by PCR to several million copies, which is well within the threshold of detection for conventional techniques (32). This has obvious uses in diagnosing genetic disorders either in carriers or prenatally. It is of interest that our most sensitive techniques at the present time for detecting protein would require about 50,000 copies of the molecule per cell, as opposed to PCR, which can amplify even a single copy of a nucleic acid to levels detectable by routine conventional techniques.

Molecular Genetics

Isolation and identification of the gene responsible for a specific inherited disorder has until recently only been possible in those diseases in which the protein defect is known. In most, neither the defect nor the responsible protein is known. Recent developments make it possible to isolate the gene despite not knowing the molecular defect. It was recognized in the late 1970s that DNA molecules on homologous chromosomes exhibit a detectable base sequence difference (polymorphism) every 300 to 500 base pairs when one allele is compared to another. These basic changes, if they occur in the recognition site of a restriction enzyme, will give rise to different size fragments than the normal allele, and this can be easily recognized. This is the basis of the restriction fragment length polymorphism (RFLP) technique. These polymorphisms, recognized by RFLP, can provide landmarks along the chromosomal DNA to which other particular markers can be linked, including a disease locus. The identification of polymorphism by restriction nucleases in the DNA of family pedigrees with members affected with a specific diseases makes it possible to link the locus responsible for the disease to markers of known chromosomal loci. This is referred to as linkage analysis based on RFLP (33). If the locus of a disease is within 10 million base pairs of the locus of a known marker, the chance of the loci separating by crossover (recombination) during meiosis is small (recombination rate of 5% to 10%), and the recombination rate decreases almost linearly in relation to the decrease in the distance between the marker and the disease. Once a disease is linked to a marker of known chromosomal locus, it follows that the chromosome carrying the disease is known, as well as its approximate site on the chromosome. If enough such markers of known chromosomal loci are available, it should be possible to find one that is in close enough physical proximity to the locus of the disease of interest that it would be genetically linked. Once the disease has been mapped to a chromosomal locus, one attempts to develop markers that flank the disease locus in as close proximity as possible. Having established close flanking markers, it is then possible to do chromosomal walking and sequence the whole region, including the gene. There are now more than 3,000 DNA markers of known chromosomal loci that can be identified throughout the human genome on the basis of minor difference in the sequences between homologous alleles. Of these, 2,144 represent genes, of which 436 are known to be associated with disease. A total of just over 6 million base pairs of the human genome has now been sequenced which, unfortunately, is less than two ten-thousandths of the total genome of 3 billion base pairs.

There are over 500 known diseases that are inherited, of which more than 50 affect the heart. Considerable progress has been made in the past year by Seidman et al. (34,35) and others (36) in attempts to isolate the gene responsible for hypertrophic cardiomyopathy. A group of Japanese investigators (37) have claimed the locus for this disease to be on chromosome 18, whereas Italian investigators claimed it to be on chromosome 16 (38). Seidman et al. (34) in two families, and our own studies in nine North American families (39) geographically distributed throughout the United States and Canada, showed the disease to be linked to the same locus, namely 14q1. Thus, it would appear chromosome 14q1 is a major locus for the disease in North America (40). It would appear that the gene responsible for the disease in these families that link to chromosome 14 is due to a mutant form of the myosin gene, however, the causal link between the myosin mutant and the disease has yet to be determined (39,41). The isolation of this gene and determination of the biochemical defect will not only help us to understand, diagnose, and treat this disorder but should also provide a clue to the fundamental aspects of cardiac growth and hypertrophy. Gene replacement therapy, while still in the future for human disease, is clearly no longer fiction but likely to be around the corner. In the case of several cardiac disorders referred to collectively as cardiomyopathies, it is unlikely that we will shed any new light on these disorders unless we apply the techniques of molecular genetics; otherwise our present attempts will remain feeble and descriptive. The clinician will play a major role in attempts to isolate the gene responsible for hereditary disease, since linkage analysis is only possible with family pedigrees and the analysis is only as accurate as the diagnosis. The application of molecular genetics demands that the investigator at the molecular level and the clinician investigating the families work as a team; otherwise the defect responsible for these disorders will remain unknown and the potential for specific therapy unexplored.

Cardiac Growth, the New Frontier

Our conceptual framework and pragmatic assessment of cardiac function for most of this century has been dominated by the biophysics of muscle mechanics, pivoted by Starling's law of the heart (42). The heart from beat to beat regulates its cardiac output in response to muscle stretch, and so preload, afterload, and contractility reign supreme. In the 1960s we recognized another important level of adaptation that coupled excitation to contractility, which emphasized the important role of ionic and hormonal regulation. While these mechanisms for adaptation are essential in both health and disease, they clearly represent rapid response mechanisms for immediate and short-term adaptation. Another important adaptive mechanism that responds quickly but is sustained and is the primary long-term response is that of compensatory cardiac growth. Compensatory cardiac growth, which pathologically is often referred to as hypertrophy, occurs in response to almost all forms of injury, including myocardial infarction, hypertension, valvular disease, cardiomyopathy, and congenital malformations. This response is not just an increase in adult muscle proteins but is also associated with the reexpression of genes active during the embryonic and fetal stages but long since suppressed.

An understanding of the mechanisms involved in cardiac growth and development of a means to modulate this process will require an intense application of the techniques of recombinant DNA and molecular biology. These efforts have been significantly stymied by the lack of a continuous cardiac cell line. The recent development of a cell line derived from atrial myocytes should markedly accelerate our progress in understanding cardiac growth. Nevertheless, considerable progress has been made using myocytes in primary culture, pressure overload, and induced hypertrophy in animals and in transgenic animals. Studies have shown that, during hypertrophy such as that induced by pressure overload in the intact animal, there is reexpression of several proto-oncogenes that clearly qualify as growth factors, which may play a role in the growth response (43,44). Several of the major questions regarding cardiac growth are being intensely explored and considerable elucidation is forthcoming. Our laboratory and others (45,46) have induced in primary cultured cardiac myocytes a growth response that is virtually identical to that seen in the intact pressure-overloaded animal, consisting of increased protein synthesis, reexpression of fetal genes, and reexpression of several proto-oncogenes. This response can be induced by adrenergic stimulation and by several known growth factors, including TGFβ, αFGF, and βFGF. The mechanism responsible for the altered gene expression can now be explored at the DNA regulatory level in the cell culture model (47).

Many important questions remain to be answered, such as what is the signal that transduces pressure into cardiac growth, what are the proteins that mediate the signal from the receptor at the cell surface to the nucleus and in what way are transcription, translation, or posttranslation events regulated in inducing protein synthesis of a select nature? However, it is already evident that the cardiac growth response is likely to come under significant inducement that will be of direct benefit in modulating this important, long-term adaptation of the heart (48). The use of gene transfer techniques in cell culture and similar experiments in the intact animal, such as with the transgenic mouse or homologous recombination, are likely to provide very important information that is fundamental to cardiac growth as well as fundamental to our understanding of this important adaptation to cardiac injury. It is unlikely that we can induce the cardiac myocyte to proliferate in the near future, but the ability to induce and modulate cardiac growth selectively is a reasonable goal.

References

1. Roberts R: Integrated program for the training of cardiovascular fellows in molecular biology. In Albertini A, Lenfant C, Paoletti R (eds): *Biotechnology in Clinical Medicine*. New York: Raven Press, 1987, pp 99–105.
2. Roberts R: Impact for molecular biology in cardiology. *Am J Physiol* 1991; 261(suppl):8–14.

3. Watson, JD, Tooze J, Kurtz DT: *Recombinant DNA: A Short Course.* New York: WH Freeman and Co, 1983, pp 242–248.
4. Avery OT, MacLeod DM, MacCarty M: Studies on the chemical nature of the substance inducing transformation of pneumococcal types. *J Exp Med* 1944; 79:137–158.
5. Watson JD, Crick FHC: Molecular structure of nucleid acids: A structure for deoxyribose nucleic acid. *Nature* 1953;171:737–738.
6. Watson JD, Crick FHC: Genetical implications of the structure of deoxyribonucleic acid. *Nature* 1953;171:964–967.
7. Franklin RE, Gosling RG: Molecular configuration in sodium thymonucleat. *Nature* 1953;171:740–741.
8. Wilkins MHF, Stokes AR, Wilson HR: Molecular structure of deoxypentose nucleic acids. *Nature* 1953;171:748–749.
9. Kornberg A: *DNA Replication.* San Francisco: WH Freeman and Company, 1980.
10. Marmur J, Lane L: Strand separation and specific recombination in deoxyribonucleic acids: Biological studies. *Proc Natl Acad Sci USA* 1960;46:453–461.
11. Doty P, Marmur J, Eigner J, Schildkraut C: Strand separation and specific recombination in deoxyribonucleic acids: Physical chemical studies. *Proc Natl Acad Sci USA* 1960;46:461–476.
12. Nirenberg MW, Matthaei JH: The dependence of cell-free protein synthesis in *E. coli* upon naturally occurring or synthetic polyribonucleotides. *Proc Natl Acad Sci USA* 1961;47:1588–1602.
13. Nishimura S, Jones DS, Khorana HG: The *in vitro* synthesis of a copolypeptide containing two amino acids in alternating sequence dependent upon a DNA-like polymer containing two nucleotides in alternating sequence. *J Mol Biol* 1981;146:1–21.
14. Olivera BM, Hall ZW, Lehman IR: Enzymatic joining of polynucleotides. V. A DNA adenylate intermediate in the polynucleotide joining reaction. *Proc Natl Acad Sci USA* 1968;61:237–244.
15. Linn S, Arbor W: Host specialty of DNA produced by *Escherichia coli.* X: *In vitro* restriction of phage fd replicative form. *Proc Natl Acad Sci USA* 1968; 59:1300–1306.
16. Smith HO, Wilcox KW: A restriction enzyme from *Haemophilus influenzae.* I: Purification and general properties. *J Mol Biol* 1970;51:379–391.
17. Kelly TJ Jr, Smith HO: A restriction enzyme from *Haemophilus influenzae.* II: Base sequence of the recognition site. *J Mol Biol* 1970;51:393–409.
18. Baltimore D: Viral RNA-dependent DNA polymerase. *Nature* 1970;226:1209–1211.
19. Temin HM, Mizutani S: Viral RNA-dependent DNA polymerase. *Nature* 1970; 225:1211–1213.
20. Cohen S, Chang A, Boyer H, Helling R: Construction of biological functional bacterial plasmids *in vitro. Proc Natl Acad Sci USA* 1973;70:3240–3244.
21. Danna K, Nathans D: Specific cleavage of simian virus 40 DNA by restriction endonculease of *Haemophilus influenzae. Proc Natl Acad Sci USA* 1971; 68:2913–2917.
22. Sanger F, Coulson AR: A rapid method for determining sequences in DNA by primed synthesis and DNA polymerase. *J Mol Biol* 1975;94:444–448.

23. Sanger F, Air M, Barrel BG, et al.: Nucleotide sequence of bacteriophage ; gFX174. *Nature* 1977;265:687–695.
24. Maxam AM, Gilbert W: A new method of sequencing DNA. *Proc Natl Acad Sci USA* 1977;74:560–564.
25. Roberts R, Grace AM: Purification of mitochondrial creatine kinase: Biochemical and immunological characterization. *J Biol Chem* 1980;225:2870–2877.
26. Basson CT, Grace AM, Roberts R: Enzyme kinetics of a highly purified mitochondrial creatine kinase in comparison with cytosolic forms. *Mol Cell Biochem* 1985;67:151–159.
27. Gething M-J, Adler B, Boose J-A, et al.: Variants of human tissue-type plasminogen activator that lack specific structural domains of the heavy chain. *EMBO J* 1988;7:2731–2740.
28. Haber E, Quertermous T, Matsueda GR, Runge MS: Innovative approaches to plasminogen activator therapy. *Science* 1989;242:51–56.
29. Jackson CV, Crow VG, Crdaft TJ, et al.: Thrombolytic activity of a novel plasminogen activator, LY210825, compared with recombinant tissue-type plasminogen activatory in a canine model of coronary artery thrombosis. *Circulation* 1990;82:930–940.
30. Puleo PR, Khatib R, Barientes S, Atmar R, Ma T: Detection of coxsackieviral infected hearts by polymerase chain reaction. *Circulation* 1990;82(suppl III):725.
31. Jin O, Sole MJ, Butany JW, et al.: Detection of enterovirus RNA in myocardial biopsies from patients with myocarditis and cardiomyopathy using gene amplification by polymerase chain reaction. *Circulation* 1990;82:8–16.
32. Ma TS, Brink PA, Roberts R, Perryman MB: A novel method to quantify low abundance mRNA in human heart using polymerase chain reaction. *Circulation* 1990;82(suppl III):50.
33. Botstein D, White RL, Skolnick M, Davis RW: Construction of a genetic linkage map in man using restriction fragment length polymorphisms. *Am J Hum Genet* 1980;32:314–331.
34. Jarcho JA, Mikema W, Pare JAP, et al.: Mapping a gene for familial hypertrophic cardiomyopathy to chromosome 14q1. *N Engl J Med* 1989;321:1372–1378.
35. Geisterfer-Lowrance AAT, Kass S, Tanigawa G, et al.: A molecular basis for familial hypertrophic cardiomyopathy. A β cardiac myosin heavy chain gene missense mutation. *Cell* 1990;62:999–1006.
36. Towbin JA, Brink PA, Fink D, Hill R, Hejtmancik JF, Roberts R: Hypertrophic cardiomyopathy: Molecular genetic exclusion of HLA linkage. *Clin Res* 1989; 37:302A.
37. Nishi H, Kimura A, Sasaki M, et al.: Localization of the gene for hypertrophic cardiomyopathy on chromosome 18q. *Circulation* 1989;80(suppl II):457.
38. Ambrosihi M, Ferraro M, Reale A: Cytogenetic study in familial hypertrophic cardiomyopathy: Indentification of a new fragile site on human chromosome 16. *Circulation* 1989;80(suppl II): 458.
39. Hejtmancik JF, Brink PA, Towbin J, et al.: Localization of the gene for familial hypertrophic cardiomyopathy to chromosome 14q1 in a diverse American population. *Circulation* 1991;83:1592–1597.

40. Roberts R (ed): *Molecular Biology of the Cardiovascular System*. Hamden, CT: Blackwell Scientific Publications, Inc (in press).
41. Tanigawa G, Jarcho JA, Kass S, et al.: A molecular basis for familial hypertrophic cardiomyopathy: An α/β cardiac myosin heavy chain hybrid gene. *Cell* 1990;62:991–998.
42. Katz AM: Molecular biology in cardiology, a paradigmatic shift. *J Mol Cell Cardiol* 1988;20:355.
43. Mulvagh SL, Roberts R, Schneider MD: Cellular oncogenes in cardiovascular disease. *J Mol Cell Cardiol* 1988;20:657–662.
44. Mulvagh SL, Michael LH, Perryman MB, Roberts R, Schneider MD: A hemodynamic load *in vivo* induces cardiac expression of the cellular oncogene, c-myc. *Biochem Biophys Res Commun* 1987;147:627–636.
45. Parker TG, Packer SE, Schneider MD: Peptide growth factors can provide "fetal" contractile protein gene expression in rat cardiac myocytes. *J Clin Invest* 1990;85:507–514.
46. Bishopric N, Simpson PC, Ordahl CP: Induction of the skeletal α-actin gene in α_1-adrenoreceptor-mediated hypertrophy of rat cardiac myocytes. *J Clin Invest* 1987;80:1194–1199.
47. Parker TG, Chow KL, Schwartz RJ, Schneider MD: Differential regulation of skeletal α-actin transcription in cardiac muscle by two fibroblast growth factors. *Proc Natl Acad Sci USA* 1990;87:7066–7070.
48. Schneider MD, Roberts R, Parker TG: Modulation of cardiac genes by mechanical stress: The oncogene signaling hypothesis. *Mol Biol Med* 1991;8:167–183.

Chapter 2

Essentials of Nucleic Acids and Proteins

The DNA alphabet is comprised of four simple letters (A, C, G, and T) that spell out the creation and function of every biological process. These letters refer to adenine, cytosine, guanine, and thymine, respectively—the four nucleotide bases found in all DNA molecules. These bases are arranged in a simple linear sequence encoding for all the information needed to carry on specific and complex functions. In the cardiovascular system this includes activities such as the production and control of cholesterol metabolism, thrombosis, contractility, electrical conduction, hormonal and autonomic effects, and various responses to disease states such as hypertension, all of which are encoded in the DNA molecule. DNA is a double-stranded molecule (1) with the two single strands bound together by hydrogen bonds that occur between exclusive base pairs. Adenine only bonds to thymine (A═T) and guanine only bonds to cytosine (G≡C). This bonding assures 100% complementarity between the two strands (2). These two concepts, the sequence of the four-letter alphabet and the highly specific complementary nature of hydrogen base pairing between the bases A and T or C and G, account for the specificity of the DNA code (3). The specific base pairing provides the basis and the specificity for practically all of the techniques of molecular biology currently in use.

This chapter provides an introduction to how this simple DNA code, contained in genetic units called *genes*, is interpreted by different cells through the production of an intermediate molecule called messenger RNA (mRNA), which is a mirror image of the genetic code contained on the parent DNA. The mRNA acts as a template to synthesize complex proteins and other cellular constituents that we hope to understand and modify in disease states using molecular techniques. Figure 2.1 illustrates the basic flow diagram of DNA producing an RNA copy that encodes for the synthesis of proteins. Some of these proteins are structural proteins, which maintain cellular integrity, and others are enzymes, which catalyze the different chemical reactions in the cells.

A gene is a discrete region on a human chromosome that contains the specific sequence that may encode for a single, unique polypeptide

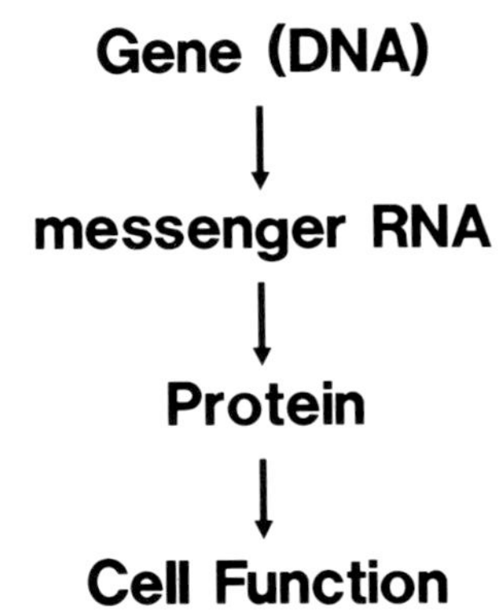

Figure 2.1 DNA controls cell function. RNA is synthesized from a gene on the DNA template in the nucleus. The protein is then synthesized from RNA to carry out cell function.

(protein). Every cell in the human body contains the genetic information to encode for the 50,000 to 100,000 types of proteins formed in the different cells. Not all of these proteins are produced in the same cell type, although some proteins are common to all cells. Each gene, therefore, must contain both protein coding sequences and regulatory sequences that allow for the functional diversity of different cell types. The concept of a gene and the elements that control its activation and subsequent gene product are outlined in the following section.

DNA and Genes

The human chromosome is the largest molecule in the human body. It is comprised of a long, double-stranded helical molecule of DNA. Genes are arranged sequentially along the length of the DNA molecule in a linear segmental fashion. Genes comprise only a fraction of the total DNA molecule. There are intervening sequences between genes that may have structural or regulatory functions. In addition, there are sequences within genes, called introns, that do not encode for the protein by may be involved in mRNA processing and editing, which is discussed in the next section.

The basic building blocks of DNA are comprised of basic chemical compounds that include 1) phosphoric acid, 2) a sugar ring called deoxyribose, and 3) four nitrogenous bases (adenine, guanine, thymine, and cytosine) (Fig. 2.2). The phosphoric acid links the deoxyribose sugars together to form a long, helical strand of DNA. Each deoxyribose sugar is linked to a single nitrogenous base. The two helical strands of DNA are connected by hydrogen bonds between the nitrogenous bases on opposite strands. One important feature is that this hydrogen binding is reversible under certain conditions. Relaxation of hydrogen binding with the exposure of a portion of the DNA as a single strand must occur for the DNA sequence (gene) to be "read" by the cell. This happens many

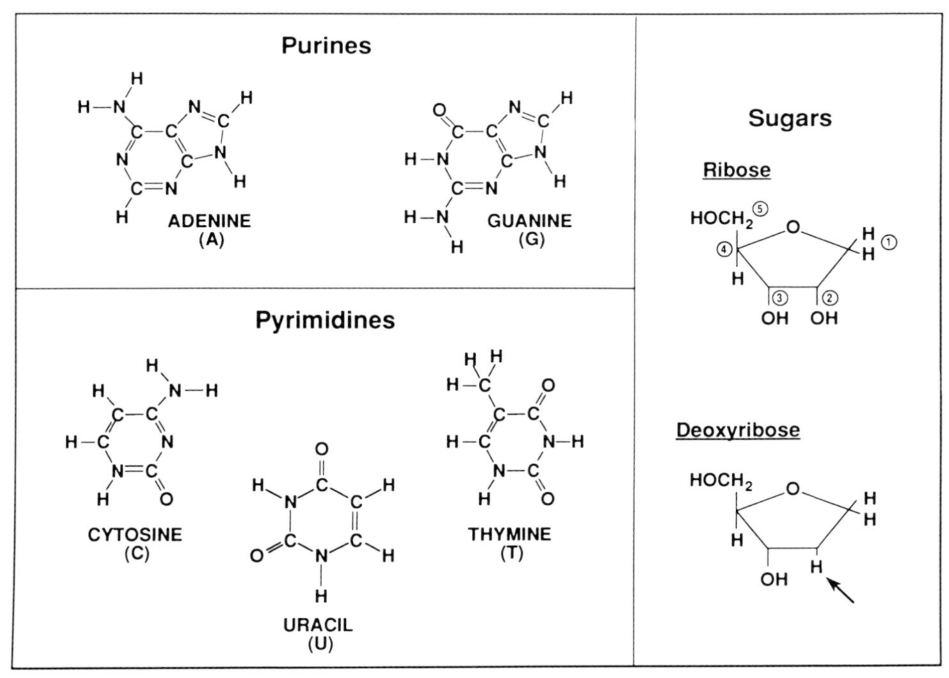

Figure 2.2 Building blocks of nucleic acids. DNA is comprised of a deoxyribose sugar backbone with the nucleotide bases adenine, guanine, cytosine, and thymidine attached to the C1 position carbon on the sugar ring.

times during a single cell cycle so that gene expression can result in normal cell function. In contrast, during cell division, the two strands of the parental double helix completely unwind and each specifies the synthesis of a new daughter strand of DNA by base pairing rules, giving two identical copies of each chromosome for each of the daughter cells (Fig. 2.3).

The DNA molecule has a polarity that determines the direction in which a particular nucleotide sequence is read. Both DNA and RNA are synthesized and read in a 5′-to-3′ direction. These numbers refer to the carbon (C) "position" on the deoxyribose and ribose sugar rings contained in these molecules (Fig. 2.4). In the case of DNA, the sugar ring utilized is the deoxyribose sugar, so called because of a hydroxyl (—OH) group missing at the C2 carbon in the ring. In contrast, RNA is synthesized using the ribose sugar ring. The specific purine (adenine and guanine) and pyrimidine (cytosine and thymine) nucleotide bases are covalently bound to the C1 carbon of the deoxyribose sugar ring. The DNA strand is formed when the ribose sugars are linked together by intervening high-energy phosphate bonds. Starting from one end of the DNA strand, the hydroxyl group on the C3 carbon of the first deoxyribose sugar is linked by a phosphate bond with the C5 carbon hydroxyl group on the next

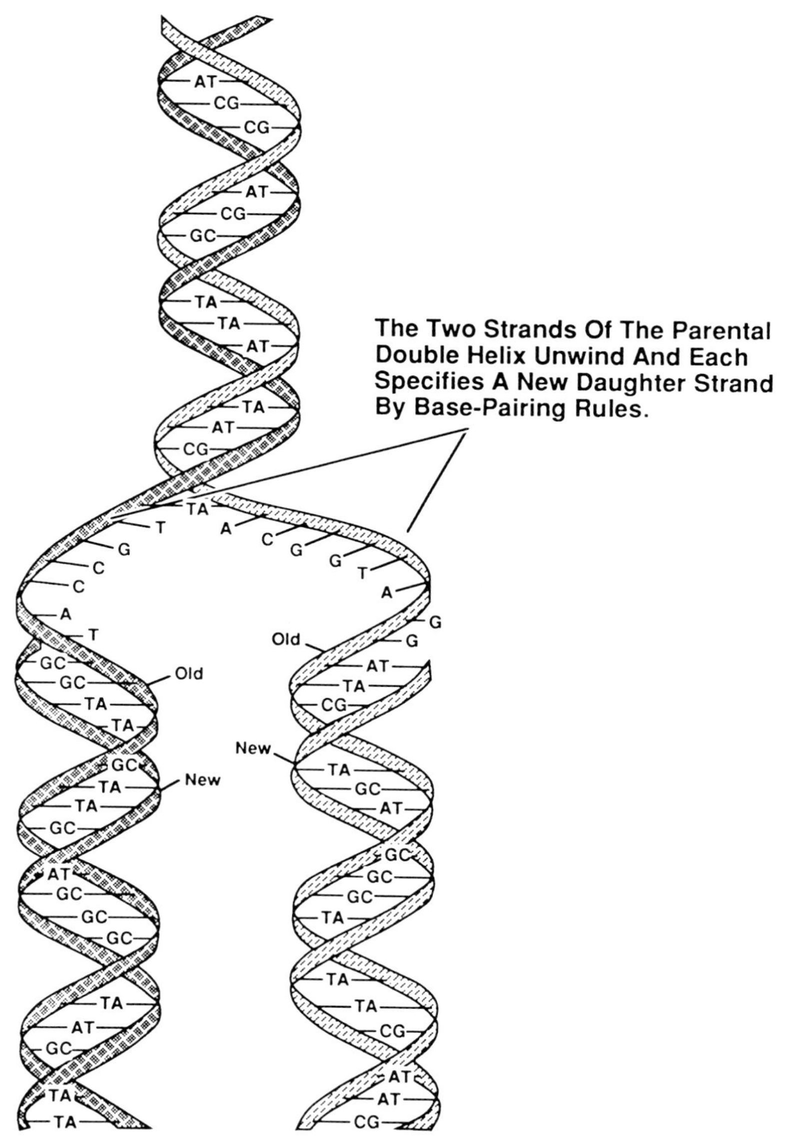

Figure 2.3 DNA replication conserves the nucleotide sequence. DNA is a double-stranded helical molecule bound together by the nucleotide bases contained on each individual strand. During cell division two identical copies of the original parental strand are made by unwinding the DNA and then synthesizing of a complementary second strand to make two identical new daughter strands.

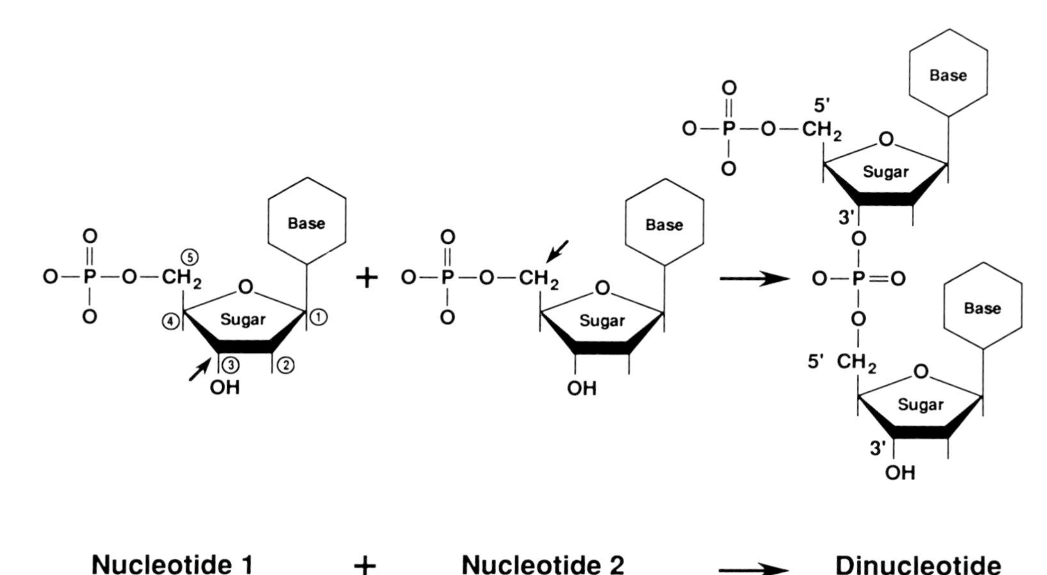

Figure 2.4 DNA is synthesized and the nucleotide sequence is read in a 5′-to-3′ direction. This refers to the carbon position on the deoxyribose sugar ring. The first nucleotide in a DNA strand has a free phosphate group on the 5′ carbon, and the strand is synthesized by adding a second nucleotide to the first nucleotide's 3′ carbon via phosphodiester bond.

deoxyribose sugar in the DNA strand, and so forth. Thus, a single strand of DNA has a free 5′ carbon hydroxyl group on one end of the strand and a free 3′ carbon hydroxyl group on the other end of the strand. The importance of this is illustrated by the example of the base sequence 5′-AGTC-3′, which is chemically different from 5′-CTGA-3′ despite the fact that they contain the same bases but with reversed sequence.

The two strands of DNA that form the double helix molecule are arranged in what is called an antiparallel arrangement. This means that one strand runs in the 5′-to-3′ direction and the complimentary strand runs in the opposite direction. The two strands are 100% complimentary, such that one strand can always predict the sequence of the opposite strand (Fig. 2.5); A always binds to T and G always binds to C. As a result, DNA always contains equimolar amounts of A and T and, similarly, the molar concentration of G is the same as that of C. However, the number of AT base pairs in a particular sequence of DNA may be different from the number of GC base pairs. This is important because the base pairs have different bonding and stability characteristics. A and T are associated with only two hydrogen bonds, whereas G and C are bound by three hydrogen bonds and, thus, are more stable than AT bonds. Therefore, regions of the double helix that are rich in AT residues can destabilize more easily than regions where GC residues are more prominent. In fact, the proportion of AT to GC within a particular DNA specimen can

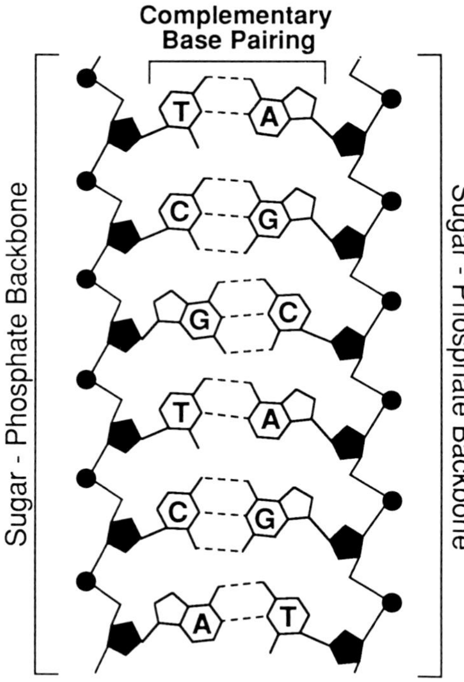

Figure 2.5 Specificity of DNA base pairing. The two strands of DNA are bound together via hydrogen bonds between the nucleotide bases on each strand. The bonds are formed by strict pairing between two complementary bases, A═T or C═G, such that each strand reflects the exact sequence of the opposite strand (*A*, adenine; *T*, thymine; *G*, guanine; *C*, cytosine). The sugar (*dark pentamer*) and phosphate (*dark circle*) linkages form the backbone of the DNA strand.

be determined by measuring the temperature required to denature the two strands. As is discussed in Chapter 3, the effects of temperature and salt concentration have been useful in laboratory analysis of DNA (4). One can adjust the stability, and thus the specificity, of binding between two strands of DNA in a predictable manner.

Gene structure, and the importance of coding and noncoding sequences within the gene, is an area of active investigation. The generic components that make up a hypothetical gene are illustrated in Figure 2.6. Most genes contain an initiation site, often defined by a structure called a *TATA box*. The TATA box, in terms of strand orientation, is at the 5′ end prior to the protein coding region of a gene. This is typically a short TA-rich sequence that is found approximately 25 base pairs before the start of transcription. Transcription is a process in which an enzyme called RNA polymerase binds to the DNA at a specific place and synthesizes a complimentary strand of mRNA base sequences from the DNA template. The synthesis of RNA always occurs in a 5′-to-3′ direction. This RNA molecule will eventually be processed and transported to the cytoplasm for protein synthesis.

The nucleotide sequences surrounding the TATA box can also influence RNA polymerase binding. These sequences may confer a tissue-specific polymerase binding that allows for expression of specific genes only in certain cell types. Collectively, the TATA box and the surrounding

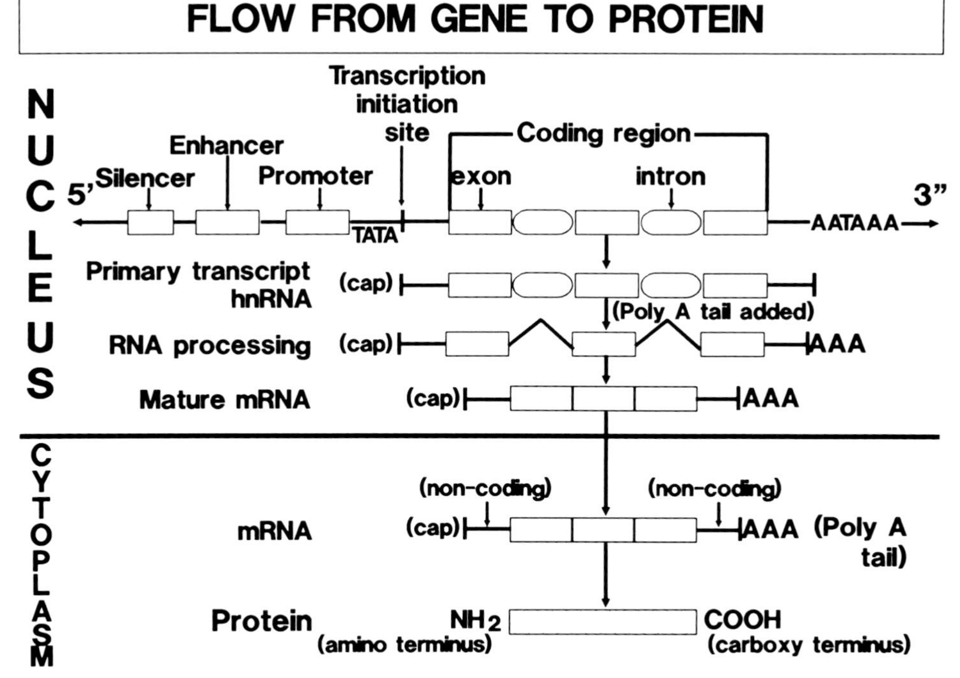

Figure 2.6 Regulation of gene activation. Schematic of the components of gene structure that contribute to gene activation and protein synthesis. RNA polymerase binds to a site at the beginning of a gene (promoter region) that often contains a TATA box. Other gene elements (enhancers) may regulate this process. A heteronuclear RNA copy of the gene (hnRNA) is produced that contains protein coding (exons) and noncoding (introns) sequences. The noncoding introns are spliced out and the RNA is "capped" at the 5′ end and polyadenylated at the 3′ end (poly[A] tail) to form a mature messenger RNA (mRNA). The mRNA is used to synthesize the protein encoded for by this gene in the cell cytoplasm.

upstream sequences are called the *promoter* of the gene (5). The promoter region is the initiation site for gene activity. While the promoter determines which genes are "turned on" in a cell, a second genetic element called an *enhancer* may be present and can modulate the promoter activity. The enhancer is a purely regulatory sequence and is not required for RNA polymerase binding. However, nuclear proteins that bind to the enhancer region sequence can have dramatic effects on the rate of RNA transcription from a particular gene. Enhancers can also function in a tissue-specific manner. Silencer sequences are regions to which proteins bind that inhibit polymerase activity and decrease transcription. Physiological stress (6), hormone receptors (7), and nuclear oncogene protein products have all been shown to have important modulatory effects on the promoter and enhancer activities for different genes (8).

Immediately downstream from the promoter region are those sequences that contain the DNA encoding for the protein product. The DNA sequences that are eventually brought together to encode for the protein are called *exons*. Within a gene, the exons are not continuous and are interrupted by intervening, noncoding sequences called *introns*. As discussed in the next section, the crude RNA transcript called heteronuclear RNA (hnRNA) contains both intron and exon sequences. Heteronuclear RNA must be processed to remove the intron sequences by splicing them out and piecing the coding exon segments together into one continuous mature mRNA molecule. Each intron begins and ends with a nucleotide sequence that is recognized by the splicing machinery, which then removes the introns and ligates the ends of the exons in tandem to make a continuous coding sequence.

Downstream from the last peptide-encoding exon of the gene is the so-called 3′ untranslated region. Although this sequence does not encode for part of the protein, it is retained in the mature mRNA transcript. One proposed function for the 3′ untranslated region is the ability to confer stability on the mRNA. This provides another possible form of gene regulation by producing an RNA transcript that is more stable, and is then more available for translation into a protein.

RNA and Transcription

Genes in the nucleus control chemical reactions in the cellular cytoplasm and extracellular fluids through intermediate nucleic acids called *ribonucleic acids*, or RNA. The formation of RNA is controlled by the DNA in the nucleus by a process called transcription. RNA is subsequently transported into the cellular cytoplasm, where it controls protein synthesis.

The RNA molecule is essentially a complimentary copy of the DNA template, from the gene from which it was transcribed. The chemical structure of RNA is essentially the same as DNA with two notable exceptions. First, ribose sugar residues that retain the C2 hydroxyl group cleaved off in the deoxyribose sugar counterpart form the strands' backbone. Second, the nucleotide base thymine is not found in RNA, and is replaced by the nucleotide base *uracil* (U), which, like thymine, forms an exclusive hydrogen bond with adenine.

Three separate types of RNA are important in protein synthesis: 1) *mRNA*, which contains the complete coding sequence for subsequent peptide synthesis; 2) *transfer RNA (tRNA)*, a unique adapter molecule that links a single specific amino acid to the appropriate position in the growing peptide chain; and 3) *ribosomal RNA (rRNA)*, which provides the scaffolding and enzymatic activity for peptide synthesis in association with mRNA and tRNA as a large RNA-protein complex (Fig. 2.7).

Messenger RNA

Messenger RNA molecules are long, single strands of several hundred to several thousand nucleotides in length, each encoding for a unique protein product. The mRNA is a "processed" form of the original hnRNA

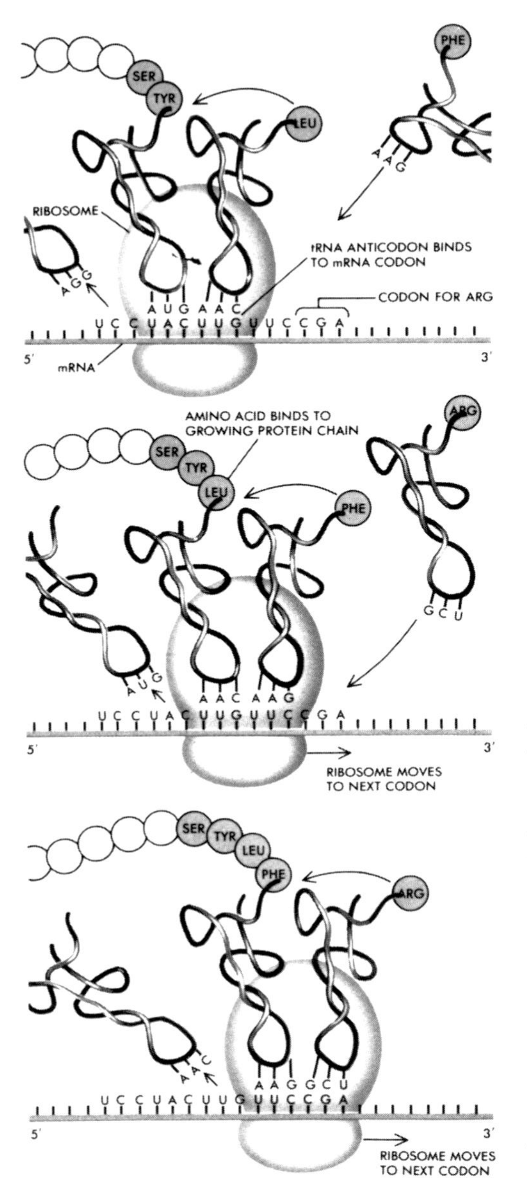

Figure 2.7 Messenger RNA directs protein synthesis. Messenger RNA binds to the ribosomal RNA complex, where its codon message is "read" by the transfer RNA anticodon sequence. Transfer RNAs, each containing their respective amino acid molecules, move on and off this large complex where peptide bonds are formed between two adjacent amino acids and protein is synthesized. The amino acid sequence of the protein is determined by the nucleic acid sequence of the RNA as it moves along the ribosome. *Reprinted with permission from Watson JD, Tooze J, Kurtz DT: Recombinant DNA: A Short Course. New York: WH Freeman and Company, 1983. Copyright © 1983 by James D. Watson, John Tooze, and David T. Kurtz.*

transcript initially transcribed from the gene (Fig. 2.6). This processing occurs in the nucleus through a specific splicing mechanism that removes the noncoding intron sequences and splices together the coding exon sequences, to form a shorter, full-coding strand of mRNA (9). Some of the exons that usually encode peptide sequences for the protein may also be spliced out in this processing event. This is an important form of gene product regulation, where inclusion or exclusion of a particular exon in a protein may produce dramatic or subtle effects on the protein's function by changing the peptide sequence. Thus, protein isoforms may

Table 2.1 The Genetic Code

1st Letter of Codon	2nd Letter of Codon				3rd Letter of Codon
	U	*C*	*A*	*G*	
U	Phe	Ser	Tyr	Cys	U
	Phe	Ser	Tyr	Cys	C
	Leu	Ser	C.T.[a]	C.T.	A
	Leu	Ser	C.T.	Trp	G
C	Leu	Pro	His	Arg	U
	Leu	Pro	His	Arg	C
	Leu	Pro	Gln	Arg	A
	Leu	Pro	Gln	Arg	G
A	Ile	Thr	Asn	Ser	U
	Ile	Thr	Asn	Ser	C
	Ile	Thr	Lyc	Arg	A
	Met	Thr	Lys	Arg	G
G	Val	Ala	Asp	Gly	U
	Val	Ala	Asp	Gly	C
	Val	Ala	Glu	Gly	A
	Val	Ala	Glu	Gly	G

[a] Codon termination.

be created that may confer an important functional diversity for a particular cell type. This process of creating protein isoform diversity by the alternate use of exons from a single gene is called *alternative splicing* (10). This mechanism is quite different from other protein isoforms, such as α- and β-myosin heavy chains, which are similar proteins produced by transcription from two separate genes (11).

The basic mRNA coding unit is called a *codon*. A codon consists of three code letters, or nucleotides, that as a unit specify for a single amino acid. The mRNA molecule is a string of consecutive codons each encoding for one of the 20 common amino acids found in protein molecules. These codons are exactly complimentary to the coding sequences found in the exons of the gene on the DNA molecule. The codon triplet also encodes for the start and stop sites for protein synthesis. Table 2.1 lists the RNA codons for the 20 common amino acids. Note that several of the amino acids are represented by more than one codon. For example, the amino acid proline can be encoded for by four different codons. Note, however, that the first two nucleotide bases in all four codons are the same, a cytosine residue. Therefore, a base substitution at the third base did not change the code. This degeneracy in the code is known as the "wobble hypothesis," which allows for several codons to encode for a single amino acid (12). The chain initiation codon AUG encodes for the amino acid methionine, which is the first amino acid in many proteins. The termination codons UAA, UAG, and UGA signal the end of protein synthesis and are found near the 3′ end of the RNA transcript.

The mRNA molecule is modified posttranscriptionally prior to transport into the cytoplasm. Two modifications occur at the mRNA ends (5′ and 3′) that appear to provide some protective feature to the mRNA, which is a relatively easily degradable molecule. The first modification is the addition of a methylated guanosine (7-methylguanosine residue) that caps the 5′ end of the mRNA transcript by a triphosphate linkage (13). The second modification is the addition of a long tail of repeated adenine nucleotides, called the poly(A) tail, to the 3′ untranslated region of the mRNA (14). In the nucleus, the poly(A) tail is approximately 250 nucleotides long and appears to confer some stability to the message. A second feature of the poly(A) tail is that, upon transport to the cytoplasm, it undergoes shortening and the final length becomes shorter and more heterogeneous for different mRNA species. It is believed that the length of the poly(A) tail confers some modulation of translation efficiency (15). As an aside, we shall see later that the poly(A) tail provides a useful chemical target for isolating the RNA transcripts from different tissues, thus allowing determination of specific gene activity in that cell type.

The importance of mRNA in terms of molecular biology techniques is that RNA isolation from a particular tissue allows one to go backward and discover which of the thousands of genes on the DNA molecule are active in that tissue. It also allows one to go forward and identify all the protein products produced by these genes, by predicting the amino acid sequence from the mRNA codons. The identities of many of these protein products remain unknown, so the identification of a unique mRNA species may be the first link to understanding some undefined structure or mechanism of function for a particular tissue or organ.

Transfer RNA

Transfer RNA is a small RNA molecule with approximately 80 nucleotides folded in a cloverleaf-like structure. It is called tRNA because it transfers single amino acid molecules to the growing polypeptide chain undergoing protein synthesis from the mRNA template. There are many different types of tRNAs, and each type will combine specifically with only one of the 20 amino acids that are incorporated into a protein molecule (16). The tRNA then acts as a carrier that transports its amino acid to the ribosome complex, where a protein molecule is being formed. In the ribosome complex each specific type of tRNA recognizes a particular single codon on the mRNA, thereby delivering the appropriate amino acid to the appropriate place in the chain of a growing protein molecule.

Before protein synthesis can begin, a specific enzyme must catalyze the linkage of an amino acid to the 3′ end of its specific tRNA molecule. Then, the tRNA molecule with its associated amino acid can find its appropriate place in the codon sequences along the length of the mRNA molecule, and the correct peptide sequence is established for the protein being synthesized. This association of tRNA with mRNA is accomplished through a triplet of nucleotide bases called an *anticodon* found at the base of the transfer RNA cloverleaf. Thus for each codon on the mRNA molecule there is a complimentary anticodon base sequence on a transfer RNA molecule. The amino acid transfer RNA complexes line up along the codons of the mRNA, with the actual recognition in binding being mediated by the transfer molecules. No contact actually exists between the amino acids and the mRNA codons. The process of peptide bond formation between the amino acids lined up on the mRNA to make a complete protein occurs by contact with the large ribosomal RNA complex, which provides both the structural organization and enzymatic activity for protein synthesis to occur (Fig. 2.7).

Ribosomal RNA

Ribosomal RNA is the third type of RNA found in the cell. Ribosomal RNA is associated with as many as 50 different types of structural and enzymatic proteins to form a large complex called the ribosome. The ribosome is composed of two physical subunits called the *40S* and *60S particles*, so called because of their different rates of sedimentation in an ultracentrifuge. The mRNA and tRNA form a complex at the 40S particle. The 60S particle is believed to provide enzymes that form the peptide bonds between successive amino acids, held in close association by the tRNA (17). The ribosome, therefore, acts as a small manufacturing plant that orchestrates the association of mRNA with tRNA and its amino acids, to allow for the synthesis of the protein molecule.

Protein Translation

The synthesis of a unique protein is the end result of the activation of a specific gene in a specific cell. This synthesized protein will then produce physiological functions in that cell, or in other cells in the case of secreted protein products. The process of converting the mRNA code into a protein is called translation (converts the nucleic acid code into an amino acid code). Protein translation can occur in the cytosol by association of the mRNA with freely floating ribosomes, or along a membranous intracellular organelle called the endoplasmic reticulum, where the ribosomes may attach during protein synthesis (18).

Protein translation occurs when a single mRNA moves along the ribosomal structure and is "read" by tRNA molecules, which move on and off the ribosomal structure as each successive amino acid is added to the chain. The message is read in much the same way as a tape is read as it passes through the head of a tape player. A single mRNA molecule is capable of synthesizing several proteins at once by associating itself with several different ribosomes at the same time. In other words, the mRNA passes through the first ribosome, and then associates with a second ribosomal structure that also begins reading its message, and so on. Ribosomes associated with the beginning of the mRNA strand (5′ end) will only have short, incomplete polypeptide chains extending from them, whereas ribosomes associated with the end of the mRNA (3′ end) molecule will have almost completed protein synthesis. Ribosomes in a cell are often seen clustered together in the cytoplasm, attached to a single mRNA. These clusters, called *polyribosomes*, are synthetic units that generate cytoplasmic proteins. In contrast, some ribosomes are bound to the endoplasmic reticulum. Protein molecules destined for secretion by the cell in secretory vesicles tend to associate themselves with the endoplasmic reticulum membrane, where they are transported into the endoplasmic reticular matrix. Ribosomes generally do not attach to the endoplasmic reticulum until these protein molecules begin to be formed. Secreted protein molecules often contain a hydrophobic tail sequence that then attracts the ribosome with its growing peptide tail to the endoplasmic reticulum membrane. Similar to polyribosomes in the cytoplasm, multiple ribosomes along the endoplasmic reticulum can process several protein molecules from a single mRNA at the same time. At the 3′ end of the mRNA molecule, a stop codon will eventually be reached, signaling the end of protein synthesis, and the mRNA molecule as well as the newly formed protein are released to the cytoplasm.

The formation of peptide bonds, like the formation of nucleotide bonds, requires energy from high-energy phosphates. Each amino acid must be activated by adenosine triphosphate (ATP) to form an adenosine monophosphate (AMP)–amino acid complex. This amino acid is now activated and can combine with its specific tRNA to form an amino acid–tRNA complex and release the AMP. On the ribosome at least two amino acid–tRNA complexes occupy adjacent codons on the mRNA. The close proximity of the two amino acids then allows for a peptide bond to be formed between them. In this chemical reaction a hydroxyl group is removed from the carboxy portion of one amino acid and the hydrogen is moved from the amino portion of the other amino acid. The two active sites of the amino acids then combine to form a covalent bond and water is released in the reaction. The energy for this reaction is supplied by another high-energy phosphate substance called guanosine triphosphate (GTP).

Gene Regulation

One of the most exciting areas of research in molecular biology is the study of regulation of gene expression. The regulation of gene expression involves the whole process from decompaction of the chromatin DNA to

Table 2.2 Regulation of Gene Expression and Protein Synthesis

Pretranscription
Transcription
Posttranscription
Translation
Posttranslation

completion of the synthesis of a mature protein. Significant control is exerted over gene expression at all levels, as indicated in Table 2.2. Very little is known about pretranscription gene regulation, wherein the chromatin DNA unfolds and exposes itself to RNA polymerase. Gene transcription means regulation of when and at what rate RNA polymerase acts. This event is considered to be the most important and probably the rate-limiting step in gene expression. Regulation of transcription is dictated by transcription factors, which are proteins that attach to the DNA. These transcription factors are proteins that bind to promoters, silencers, and enhancers (regulatory regions) at the 5′ end of the gene. There is also considerable control of the posttranscriptional processing of mRNA, whereby different protein isoforms are encoded by alternate splicing of mRNA exons. Translation of mRNA into protein may also be regulated, at the level of polypeptide assembly, by the amount of that protein already present in the cell. Posttranslation modification of proteins that targets them to different subcompartments of the cell cytoplasm, such as the cell membrane or organelles, and other protein modifications, such as glycosylation or proteolysis, occur that are necessary for the protein to perform its function.

The vast difference in cell types in the body result from activation and repression of different types of genes. Certain "early" genes regulate the differentiation of progenitor embryonic stem cells into their respective cell types. Cellular differentiation typically occurs during embryogenesis, although a small number of undifferentiated cells retain this capacity throughout life. Dedifferentiation of a differentiated cell to a relatively undifferentiated cell may also occur under certain physiological conditions, or may be an abnormal response, as seen in tumorigenesis (19).

When a functioning differentiated cell has been created by the activation and repression of certain genes, a second type of gene regulation may come into play. Genes in the cell that may be functioning at low levels may get turned on in response to physiological activation of the cell. Examples of an activation of gene responses through a physiological stimulus would be the infiltration of heart tissue with activated inflammatory cells in viral myocarditis, hypertrophic responses with contractile protein isoform changes in pressure-overloaded cardiac myocytes, and regulation of β-receptor expression in a heart with long-standing congestive heart failure (20). Both steroid and thyroid hormones are also examples of stimuli that can modulate the gene expression in their target cells (21). The implications for controlling gene regulation in medicine include the immense possibilities for genetic and pharmacological gene manipulation to correct or control human disease. The following section illustrates some of the target points for regulation of genes and their subsequent protein activity.

DNA Packaging

In humans, the 22 autosomal chromosome pairs plus the two sex chromosomes (X and Y) contain 3 billion nucleotide base pairs, all tightly packaged and organized in a small nucleus that is only about 5×10^{-6} meters in diameter. This tight packaging of the DNA is accomplished by supercoiling the DNA through its association with large complexes of several different nuclear proteins called *histones*. There are four distinct histone molecules (histones H2A, H2B, H3, and H4) that, in pairs, form an octomer of histones around which DNA is supercoiled. This supercoiled DNA-histone complex is called the *nucleosome* (22). In cells visualized by electron microscopy, these structures appear as beads on a string, separated by a thin strand of DNA that is not associated with the nucleosome. A fifth type of histone (histone H1) is known to bind to these intervening strands of DNA. As one might imagine, the availability of a gene for active transcription might depend on the degree of histone binding and DNA supercoiling in that region. It is believed that nucleosomes and supranucleosomal structures are relaxed or unfolded within transcriptionally active chromatin (Fig. 2.8). While histones are retained in the coding region of genes when they are transcriptionally active, the DNA-histone interactions within the gene appear to be altered (23). Furthermore, no histones can be detected in the promotor regions of these activated genes. This process appears to be reversible, and the exact mechanisms underlying histone binding regulation are currently under investigation.

DNA Methylation

It would not be useful to have all genes active in a single tissue type. One proposed mechanism for inactivation of a tissue-specific gene in an unrelated tissue is the process of DNA methylation (24). This involves the addition to a specific sequence in the gene (usually at the 5′ end near the promotor) of a methyl group (CH_3—) that functions by blocking transcription initiation from that gene. DNA methylation typically occurs on a cytosine residue situated immediately 5′ from a guanine residue and that is often referred to as *CG dinucleotide*. This methylated CG dinucleotide appears to cause a structural change in the DNA that blocks

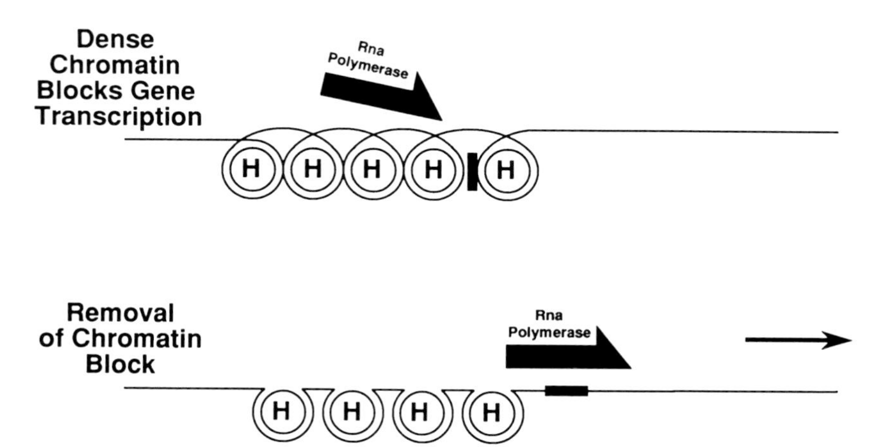

Figure 2.8 Chromatin structure affects on gene activation. Schematic representation of DNA (*line*) association with histones (*H*). A conformational change in histone binding may allow RNa polymerase to bind to the promoter (*dark rectangle*) and initiate gene transcription.

the ability of RNA polymerase II to attach to the DNA and begin transcription from that gene (25). Differentiating and developing cells appear to have tissue-specific factors that are capable of demethylating genes that are required for the development of that specific cell type. Thus, in the heart, gene activation may be accompanied by a site-specific and tissue-specific demethylation of a particular gene necessary for normal cardiac cell differentiation (Fig. 2.9).

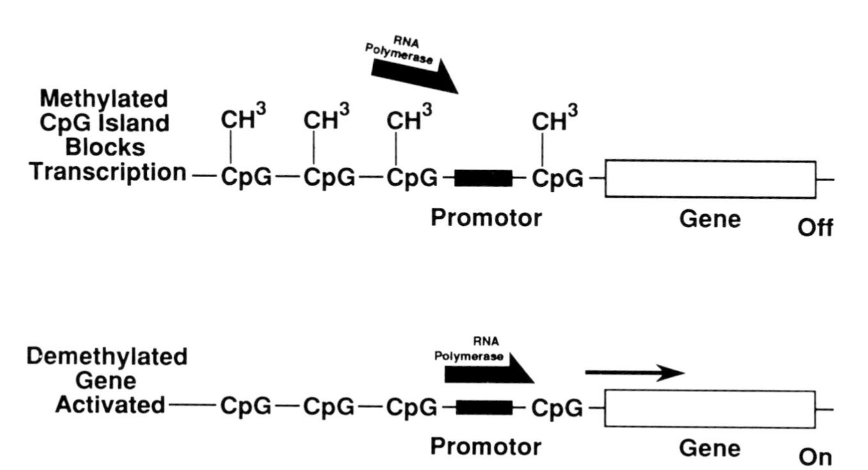

Figure 2.9 DNA methylation affects on gene activation. The presence of methyl groups (CH_3) bound to the cytosine-guanine dinucleotide near the promoter region of a gene is believed to block the activity of a gene. Unmethylated promoter regions often coincide with active genes.

Interestingly, hypomethylated CG dinucleotides may be used as a marker for identifying regions of anonymous DNA that may contain a nearby gene that remains undefined. CG dinucleotides are in general unrepresented in most DNA sequences. However, there is a tendency for CG dinucleotides to cluster at the 5′ end of the potentially active gene. This clustering of CGs at the 5′ end of a gene is often referred to as a "CG island" (26). Some investigators have taken advantage of certain enzymes that can specifically cut DNA at a CG island to help them identify an unknown gene in that region of the DNA molecule.

Transcription Factors

Gene activity is partially determined by the ability of the enzyme RNA polymerase to bind to the promoter region and begin transcription from the gene. The ability of RNA polymerase to bind to the region is inhibited or enhanced through modulation by several different types of proteins, referred to as *transcription factors*, that bind to the promoter, enhancer, and silencer sequences discussed earlier. Transcription factors and the DNA promoter to which they bind are currently an area of active research because of the potential for genetically engineered control of gene activation. These transcriptional factors are regulatory proteins that have been classified based upon their predicted secondary protein structure. There are three generally accepted structural motifs that are named on this basis: 1) helix-turn-helix motif, 2) zinc finger motif, and 3) leucine zipper motif (Fig. 2.10). Each motif is a broad category for a family of related protein that each recognize similar DNA sequences (27). The general function or specificity of these motifs is discussed next.

Helix-Turn-Helix Motif

As the name implies, the crucial structure in this class of proteins is that of α-helices separated by a β-turn. The structure can be thought of as cylindrical rods representing the α-helixes connected end to end by a string that allows a turn, such that the cylinders may be at sharp angles to each other. This structure inserts itself into the DNA by interactions of one of these helices, the so-called recognition helix, with nucleotide bases exposed in the major groove of the target DNA. This allows the other helix to lie across the groove and make other contacts to the DNA (28). Usually there are three helices, but there may be four or more.

The importance of helix-turn-helix proteins is that these molecules control many of the key decisions in early development that determine the fate of a particular cell. They can be thought of as a master switch that can activate a cascade of genes important in cell type–specific determination, such as those cells destined for neural tissue formation ver-

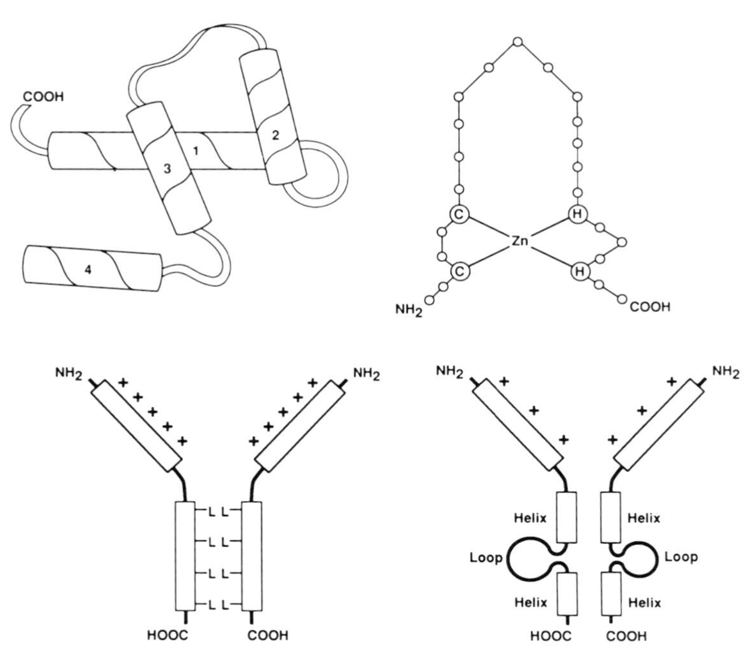

Figure 2.10 Types of transcription factors which effect gene activation. Schematic representation of the shape of four types of protein transcription factors that bond to DNA and influence gene activation. Helix-turn-helix is a protein with two α-helices separated by a β-turn. Leucine zippers are protein dimers with interdigitating leucine amino acids. Zinc fingers have a peptide loop connected at the base by a zinc ion tetrahedron between cysteine and/or histidine in amino acids. The helix-loop-helix consists of α-helix but utilizes leucine zippers and has a loop between the α-helices. The *darkened areas* are believed to be the regions of the protein that interact with the DNA to modulate transcription.

sus those cells destined to form neural embryonic limb buds. The DNA binding site for this class of proteins is called a *homeodomain* and is important in both prokaryotes and eukaryotes.

Helix-Loop-Helix Motif

The helix-loop-helix motif is utilized by a group of transcription factors important for muscle development. One of the molecules of this class is the helix-loop-helix protein called myo-D. Myo-D is a nuclear protein that is capable of turning on the gene programs that create a normal skeletal muscle cell. It does this by binding to a DNA sequence called an *E-box*, present in the promoter region of most muscle genes. The influence of this protein on gene activity is so strong that if myo-D is introduced into a nonmuscle cell, it is capable of transforming that cell into a skeletal muscle cell by imposing its regulatory properties. A homologous helix-loop-helix protein that is responsible for differentiating a progenitor cell into a cardiac muscle cell (myocyte) has not yet been established, and is an area of active investigation. The potential ability to convert a cardiac fibroblast into a cardiac myocyte has intriguing implications for future approaches to heart failure.

Zinc Finger Motif

It is well known that some hormones exert their activity by binding to a target cell and then being transported to the nucleus, where they interact directly with the DNA. Glucocorticoid and thyroid receptor proteins represent a distinct class of specific DNA binding proteins that contain a so-called zinc finger motif. A zinc finger motif is a protein that forms a fingerlike loop that is stabilized at the base by the presence of a single zinc ion, which forms a tetrahedral coordination site between four specific amino acids (29). In the case of glucocorticoid and thyroid receptors, all four amino acids are cystine residues. The importance of zinc finger proteins in the heart is illustrated by the example of thyroid hormone. Thyroid hormone binds to cell surface receptors on cardiac myocytes, and the receptor complex is transported to the nucleus, where it has been shown to be a prominent regulator of myosin heavy chain gene expression.

A second type of zinc finger motif forms the same looplike structure, except the zinc ion is bound to two cystine and two histidine amino acids at the base. For both types of zinc finger motifs, depending on the protein, the number of potential zinc fingers ranges from two to more than 10 in a single molecule. It is believed that zinc finger proteins play an essential role wherein a functional DNA binding domain for RNA polymerase is created by zinc finger association with the DNA.

Leucine Zipper Motif

Cardiac hypertrophy is an example of a physiological stress that induces the activation of typically quiescent genes and leads to alterations in cardiac myocyte morphology. It has been postulated that the regulation of this gene activation is controlled by a group of nuclear proteins called *nuclear oncogene proteins*, or oncoproteins. The names of some of these oncogene proteins include *jun* and *fos*. These oncoproteins generally form dimers that can bind to enhancer or promoter regions to stimulate RNA synthesis from a quiescent gene. Part of their ability for differential regulation comes from the fact that some stimuli may only produce certain categories of oncoproteins. Before acting, these oncoproteins bind as dimers to themselves or to another oncoprotein, thus forming either a homodimer or a heterodimer. The leucine zipper is a structural motif that allows for binding of two of these proteins together to form a functional transcriptional activator. The way they bind is dependent upon the presence of four or five leucine residues that are spaced exactly seven residues apart in the protein sequence. The complimentary stretch of leucine residues on a second oncoprotein then allows for interdigitation of the leucine groups in a zipperlike fashion, forming the functional dimer (30). The ability to form heterodimers between two different oncoproteins allows for the possibility of yielding new proteins with distinct recognition properties. This may make it possible to influence gene expression only when two specific physiological conditions occur, producing two oncoproteins that then form a single leucine zipper transcriptional activator. All of the leucine zipper motif proteins are capable of recognizing and interacting with a common sequence on the DNA, usually known as the AP-1 site. The AP-1 site is a specific, conserved DNA sequence that is only found in the regulatory region of certain genes.

Promoters and Enhancers

Promoters and enhancers can be thought of as units of modules of DNA sequences located upstream from the protein coding gene (Fig. 2.11A). The transcriptional factors discussed in the preceding section integrate with these modules, allowing different genes to evolve distinct and com-

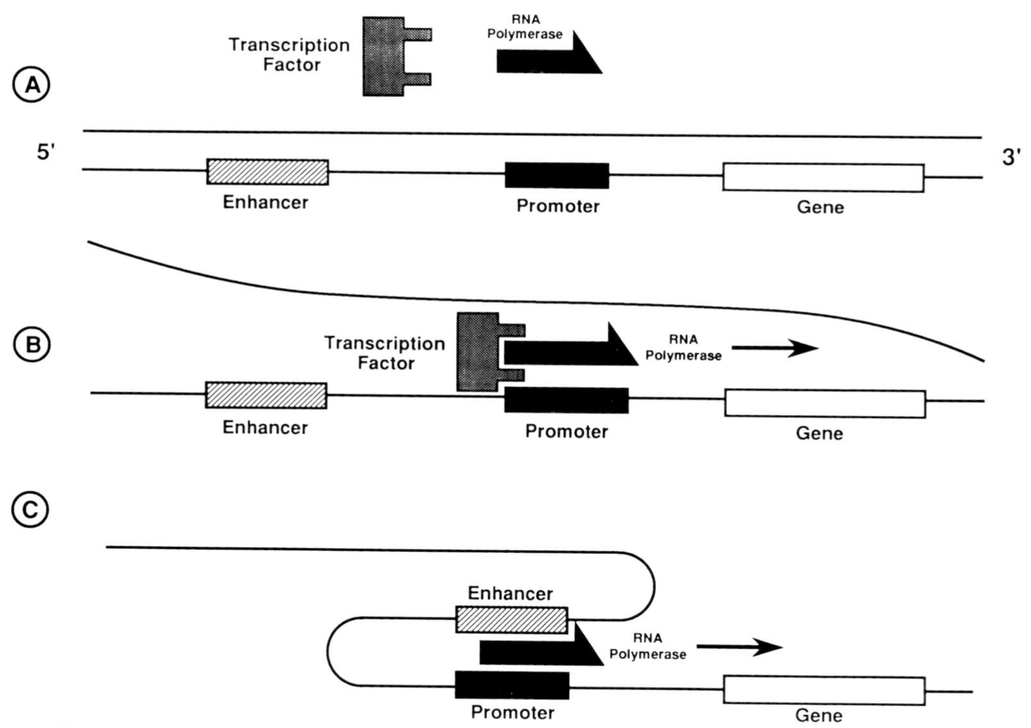

Figure 2.11 RNA polymerase binding is influenced by DNA and protein elements. Several elements interact in *cis* (promoters and enhancers) or in *trans* (transcription factors) to influence gene activity. (A) The 5′ end of a gene contains promoter and enhancer elements that can interact with transcription factors and RNA polymerase. (B) The DNA strands separate and RNA polymerase binds to the promoter in association with a transcription factor to regulate gene transcription. (C) DNA looping allows interaction of enhancer and promoter elements to modulate RNA polymerase binding and gene activation.

plex patterns of transcriptional regulation. The term *promoter* is used to refer to DNA sequences that act as binding sites for transcription factors (proteins) that form transcriptional modules clustered around the initiation site for RNA polymerase (31). At least one of these modules functions to position the start site for RNA synthesis (Fig. 2.11B). The best known example of this is the TATA box, although some promoters lack a TATA box and have other transcriptional initiation sequences. Some genes may have several promoter modules, which would thus increase the frequency of transcriptional initiation by providing several RNA polymerase binding sites. Depending on the promoter, it appears that individual modules can work either cooperatively or independently to activate transcription.

Enhancers, in contrast, are DNA sequences that form module units that bind transcriptional regulatory proteins and do not contain a transcriptional initiation site. Enhancers also have the ability of acting over a large distance of DNA (32). In other words, enhancers may be located near the promoter region or maybe at distances of 40,000 to 50,000 base pairs (bp) upstream while still maintaining the ability to enhance transcription from a particular promoter. In cases in which the promoter and enhancer for a gene are not closely linked, it is believed that DNA looping occurs to bring these regions together and allow for this long-range signaling event (Fig. 2.11C). Many promoters and enhancers are only activated by physiological stimuli such as the presence of a hormone receptor binding protein (zinc finger). Other promoters and enhancers appear to be active in all cells, such as those for "housekeeping genes," required for basal cellular metabolism.

It is interesting that some of the "regulated" promoters and enhancers for tissue-specific genes also contain modules that are capable of binding proteins that are active in all cells. Such proteins are said to be "constitutively expressed" (33). An example of a constitutively expressed transcription factor protein is one called Spl. It is believed that the presence of an Spl binding site in regulated promoter and enhancer sequences contributes to low levels of basal transcription of these genes. Alternatively, it may help to create an environment in which transcription from the gene is more highly responsive to the presence of other inducible proteins, because the gene is "primed" for activation by the presence of an Spl site.

RNA Processing

RNA processing is the next level of control in the processing of a specific gene product. As mentioned earlier, the primary hnRNA transcript is larger than mature mRNA, which encodes for protein in the cytoplasm, because noncoding introns are still interspersed between the coding exon sequences. The hnRNA transcript must be chemically modified at the ends and the introns must be spliced out before it can be transported to the cytoplasm for protein synthesis. The extent of these modifications and different cell types allows for a further mechanism of regulation of the ultimate gene product.

This section focuses on the ability of a cell to modify the final coding sequence of mRNA through an additional mRNA processing event called *alternative splicing*. Alternative splicing is simply an extension of the RNA processing events that have already been described. During the transition from hnRNA to the mature mRNA transcript, introns must be spliced out. The final exon sequence can then be brought together as a continuous mRNA coding sequence that will be "read" during protein synthesis. One example of an alternative splicing process is when selected exons that may encode for part of the protein in one tissue may be removed or spliced out in another tissue. The final amino acid sequence for this individual protein is thereby different in the two tissues. This alternative splicing event produces different isoforms of the protein product from a single gene. Some of these isoforms may be produced in a tissue-specific manner. This is one way in which a protein's function can be tailored for a specific tissue type.

In cardiac muscle as well as skeletal muscle, differential splicing has been observed in a number of different proteins (34). This list includes some of the contractile proteins such as alkaline myosin light chain, tropomyosin, and troponin-T. Some of the structural proteins in skeletal and cardiac muscle are also alternatively spliced. Two cytoskeletal proteins, ankyrin and dystrophin, which are known to bind membrane receptors, are alternatively spliced in a tissue-specific format (35,36). A third cytoskeletal protein, spectrin, as well as a number of sodium, potassium and calcium ion channels, have been shown to exhibit a number of isoforms (37) that are probably formed by a similar mechanism of differential splicing. The molecular details of some of these interesting protein isoforms are discussed in Chapters 5 and 7.

Figure 2.12 illustrates several different recognized patterns for alternative splicing. The most common types of splicing include those that change the amino acid sequence in a specific region of the protein, usually those sequences involved in protein activity. Cassette-type splicing involves the inclusion or exclusion of single or multiple exons in a particular tissue. Mutually exclusive splicing refers to a situation in which only one of two adjacent exons can be present in the final transcript. If one exon is included in the transcript, the other exon will automatically be removed or spliced out. Sometimes sequences that are removed as intron sequences in one tissue may be found included in the final transcript in another tissue, adding a few amino acids to the middle of the protein. In other cases, the retained intron material may cause the RNA polymerase to recognize an early stop codon contained within the intron.

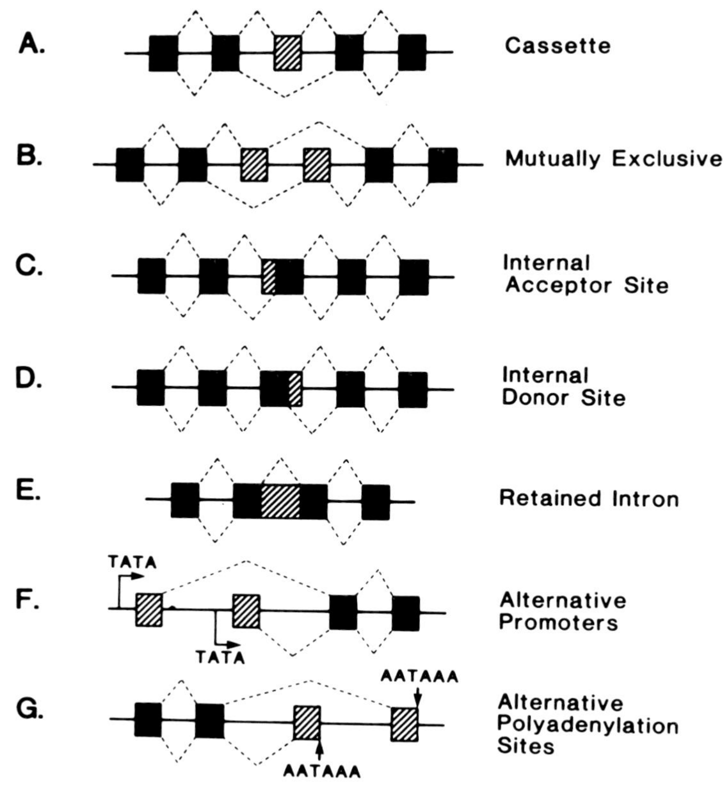

Figure 2.12 Alternative splicing of mRNA. Constitutive exons (*black*), alternative sequences (*striped*), and introns (*solid lines*) are spliced according to different pathways (*dotted lines*), as described in the text. Alternative promoters (TATA) and polyadenylation signals (AATAAA) are indicated. *Reprinted by permission from Breitbant RE, Andreadis A, Nadal-Ginard B: Annu Rev Biochem 56:467, 1987.*

This would cause early chain termination and result in an unstable or nonfunctional protein. Alternative splicing creates considerable diversity in the possible protein products from a single gene, and provides another mechanism for cell type–specific function.

Noncoding Genomic DNA

While much of this discussion has been centered around the coding regions of DNA called genes, it should be kept in mind that this comprises only a small fraction of the total DNA found in the human genome, probably less than 5%. The function of the remaining intervening regions of noncoding DNA between the genes is not well understood. As was mentioned previously, some of these DNA segments are involved in histone binding and perhaps regulation of gene function. Other portions of this interspersed DNA contain different "families" of repetitive sequences that have been copied throughout the genome many times over. In fact, at least 20% to 30% of the human genome consists of repetitive sequences, which have now been subclassified into different types (38). This section discusses several of the repetitive element families that are dispersed throughout the genome. Their functions remain to be elucidated, but their predictable locations along individual chromosomes have provided useful markers for genetic studies. For example, these sequences have been used to isolate specific pieces of DNA from a complex mixture, and have also been useful in identifying genetic markers for following the inheritance of human disease.

SINE (Alu) Repeats

One family of repetitive DNA elements is called SINES, or short interspersed repetitive elements. SINES are relatively short sequences that range in length from 130 to 300 bp and may be present in many thousands of copies per genome. The most abundant of the SINES is the so-called *Alu family* of repeats (39). The human *Alu* sequence is a 300-bp sequence that derives its name from the presence of a cleavage site specific for the restriction enzyme *Alu*I. Restriction enzymes (which cut DNA at defined sequences) are discussed in Chapter 3. It has been estimated that there are approximately 3×10^5 copies of the *Alu* repeat distributed throughout the human genome. Although the distribution is not uniform, the *Alu* repeats occur approximately every 5,000 bp. The positions of the *Alu* sequences in the human genome are highly conserved.

The reiterative number of copies of *Alu* repeats suggest that these are "mobile" units of DNA. It has been hypothesized that *Alu* sequences are transcribed in a reverse transcription event, making a complimentary *Alu* DNA sequence (cDNA) allowing for reinsertion of this *Alu* in another portion of the DNA. Sequences that undergo reverse transcription and reinsertion into the DNA are called *retroposons*. Retroposons are one form of mobile DNA. In some cases, evolutionary transmission of a segment of DNA from apes to man can be followed by retroposon insertion of an *Alu* sequence into the DNA segment in a pattern consistent with other evolutionary markers. Thus, the divergence of *Alu* insertions across species has been used to determine patterns of evolutionary inheritance.

The human *Alu* sequence consists of two repeated sequences. One sequence is 130 bp, and the second sequence is 130 bp with a 40-bp direct repeat inserted in the sequence, making its length approximately 170 bp. Thus, the entire *Alu* repeat consists of tandomly arranged 130-bp and 170-bp sequences and is approximately 300 bp in length. The *Alu* segment is flanked by short, direct sequence repeats of 7 to 10 bp, which is typical of genomic gene elements that have been reinserted into the chromosome at another location. In addition, the 3′ end of the *Alu* sequence has traces of a poly(A) region reminiscent of the poly(A) stretches found in normally transcribed mRNA. These multiple factors are evidence for the retroposon theory of the *Alu* family evolution. While the function of the *Alu* sequence is not known, it is found throughout the genome both within genes and in noncoding sequences of DNA. The sequence is about 80% homologous to a small subclass of very common cytoplasmic RNAs that are approximately 7S in size. The function of these small RNA molecules is also not well understood, but they appear to be associated with the 60S subunit on the ribosome and are involved in the processing of leader sequences present on secreted proteins when they are synthesized. The sequence homology suggests that the gene that encodes for 7S RNA may be the distant ancestor from which *Alu* repeats originated (40).

As discussed in Chapter 3, the high frequency of *Alu* in the human genome provides a convenient marker for several laboratory techniques in molecular biology. For example, if an *Alu* sequence of DNA is radioactively labeled and then added to a complex DNA mixture, other pieces of DNA that contain the *Alu* element can be easily identified. The complementary base sequences between *Alu* repeats allow the radioactive *Alu* sequence to bind in a complementary manner to the *Alu* sequences found in the unknown portion of DNA. This and other methods using the *Alu* repeat family can be useful in cloning experiments to "fish out" human DNA containing *Alu* from the background of nonhuman DNA present in hybrid cells, bacteria, yeast, and viral phage cloning systems (used to propagate the human DNA) (41).

LINE Repeats

The family of LINE, or long interspersed repetitive elements, repeats represents a larger (approximately 5,000-bp) and less frequent repetitive element present at more than 10^4 copies per genome. These repetitive elements are present less frequently than the *Alu* elements, and occur approximately every 10,000 to 15,000 bp along the chromosome. Synonyms for this family of repeats include L1 or Kpn1 repeat elements. LINE sequences also have flanking repeats and poly(A) tails typical of retroposons, or "mobile" DNA. The function of these large repetitive elements, like that of the *Alu* family, is not well understood. They also may be used as important marker sequences along the DNA. In contrast to *Alu* sequences, there has been some speculation that LINE sequences are transcribed into mRNA, which actually encodes for a protein that may influence DNA transcription. LINE sequences may be found within introns of genes as well as in noncoding DNA segments (38).

Donehower Sequence

This relatively small, 70-bp repetitive sequence was first described by Donehower et al. (42). It is a non-*Alu* sequence that is not believed to be a retroposon. The sequence has been identified in more than 30 human genes and is located in a conserved position between man and mouse. In many of these cases it was initially hypothesized that these repeat elements maybe localized specifically to regions of DNA that contain genes. However, large sequencing projects, such as 57,000 bp sequenced around the human hypoxanthine phosphoribosyltransferase (HPRT) locus of man, have shown that, in addition to being located within the gene, Donehower sequences are found flanking the locus in the noncoding segments. If this finding is extended to other intergenic regions, this repeat may not be localized specifically to genes.

Short Tandem Repeats

Short tandem repeats (STRs) are, as their name implies, short runs of dimeric (ACACAC . . .), trimeric (AGCAGCAGC . . .), or tetrameric (AATCAATCAATC . . .) sequences that are found throughout the human genome. These nonretroposon sequences are found to have stably inheritable chromosomal positions (loci) and sequence lengths within families (43). Their importance lies in the fact that their length is often highly variable in unrelated individuals. This variability between individuals (called *polymorphism*) is extremely useful when applied to genetic studies. An STR sequence that is found to be consistently transmitted in those who have the disease in a family with an inherited disease provides a reliable disease marker in that family. Besides being useful for prenatal DNA diagnosis, these sequences may help identify a region of a chromosome that contains this gene and eventually lead to isolation and characterization of the gene and its product. We discuss the importance and application of *polymorphic markers* in molecular genetics in Chapter 4.

Pseudogenes

Pseudogenes are not repetitive elements but are included in this section because of their proposed origin, which is a retroposon, or reversed transcription of the mRNA of the original normal gene sequence. Pseudogenes are nonfunctional genes found in the genome that are homologous to the normally expressed genes but often located on a different chromosome. Pseudogenes usually contain an alteration in their sequence that either changes the reading frame to a nonsensical message or encodes for an early stop codon, creating a nonfunctional gene product. Pseudogenes do not contain the intron sequences found in the normal gene and are usually very similar to the processed mRNA sequence produced from the functional gene. It is suspected that these intronless genes arise through the rare reverse transcription process of cytoplasmic mRNA, creating a complimentary DNA (cDNA) that is then somehow randomly inserted into the genome (44). One possible source for the reverse transcription enzyme that produces a cDNA from a mature mRNA is the infection of an organism with a retrovirus that contains this enzyme in order to propagate its own genome. For such reinserted sequences to be inherited, they must be present in the progenitor cells that give rise to germ cell lines (ova or sperm) and must be propagated through subsequent progeny. If such a reversed transcription event were to occur in a somatic cell, there would be no hereditary consequences.

Pseudogenes can be important from a molecular analysis standpoint. RNA sequences used to find the original gene segment on the human chromosome are perfectly complimentary to the exon segments contained in the original gene. However, the presence of a pseudogene that is nonfunctional but contains similar complimentary sequences may confound the analysis by allowing isolation of the pseudogene rather than the desired functional gene from the random mixture of DNA. This problem is often easily circumvented by sequence analysis. The formation and positioning of the pseudogenes are of interest for some investigators.

References

1. Watson JD, Crick FHC: Molecular structure of nucleic acids: A structure for deoxyribose nucleic acid. *Nature* 1953;171:738–738.
2. Crick FHD, Watson JD: The complementary structure of deoxyribonucleic acid. *Proc R Soc [A]* 1954;223:80–96.
3. Crick FHC, Barnett L, Brenner S, Watts-Tobin RJ: General nature of the genetic codes for proteins. *Nature* 1961;192:1227–1232.
4. Wetmur JG: Hybridization and renaturation kinetics of nucleic acids. *Annu Res Biophys Bioeng* 1976;5:337.
5. Youderian P, Bouvier S, Susskind M: Sequence determinants of promoter activity. *Cell* 1982;30:843–853.
6. Izumo S, Jompre AM, Matsuoka R, et al.: Myosin heavy chain messenger RNA and protein isoform transitions during cardiac hypertrophy. *J Clin Invest* 1987;79:970–977.
7. Lompre A-M, Nadal-Ginard B, Mahdavi V: Expression of the cardiac ventricular α- and β-myosin heavy chain genes is developmentally and normally regulated. *J Biol Chem* 1984;259:6437–6446.
8. Komro I, Kurabayashi M, Takaku F, Yasaki Y: Expression of cellular oncogenes in the mytochondriun during the development stage and pressure-overloaded hypertrophy of the rat heart. *Circ Res* 1988;62:1075–1079.
9. Leff S, Rosenfeld M, Evans R: Complex transcriptional units: Diversity in gene expression by alternative RNA processing. *Annu Rev Biochem* 1986; 55:1091–1117.
10. Andreadis A, Gallego ME, Nadal-Ginard B: Generation of protein isoform

diversity by alternative-splicing: Mechanistic and biologic implications. *Annu Rev Cell Biol* 1987;3:207–242.

11. Mahdavi V, Chambers AP, Nadal-Ginard B: Cardiac α- and β-myosin heavy chain genes are organized in tandem. *Proc Natl Acad Sci USA* 1984;81:2626–2630.
12. Crick FHC: Codon-anticodon pairing: The wobble hypothesis. *J Mol Biol* 1966; 19:548–555.
13. Shatkin AJ: Capping of eucaryotic mRNAs. *Cell* 1976;9:645–653.
14. Fitzgerald M, Shenk T: The sequence 5′-AAUAAA-3′ forms part of the recognition site for polyadenylation of late SV40 mRNAs. *Cell* 1981;24:251–260.
15. Jackson RJ, Standard N: Do the poly(A) tail and 3′ untranslated region control mRNA translation? *Cell* 1990;62:15–24.
16. Ikemura T: Correlation between the abundance of *Escherichia coli* transfer RNA's and the occurrence of the respective codons in its protein genes. *J Mol Biol* 1981;146:1–21.
17. Chambliss G, Craven GR, Davies J, et al (eds): *Ribosomes: Structure, Function and Genetics*. Baltimore: University Park Press, 1980.
18. Alberts BE, Bray D, Lewis J, et al (eds): *Molecular Biology of the Cell*. New York: Garland Publishing, Inc, 1983.
19. Blau HM, et al: Plasticity of the differentiated state. *Science* 1984;230:758–766.
20. Bristow MR, Ginsburg R, Minobe W, et al: Decreased catecholamine sensitivity and β-adrenergic-receptor density in failing human hearts. *N Engl J med* 1982;307;205.
21. Evans RM: The steroid and thyroid hormone receptor super family. *Science* 1988;240:889–895.
22. Richmond T-J, Finch JT, Rushton B, et al: Structure of the nucleosome core particle at 7 A resolution. *Nature* 1984;311:532–537.
23. Nacheva GA, Guschin DY, Preobrazhenskaya OV, et al: Change in the pattern of histone binding to DNA upon transcriptional activation. *Cell* 1989;48:27–36.
24. Riggs AD: X inactivation, differentiation, and DNA methylation. *Cytogenet Cell Genet* 1975;14:9–25.
25. Razin A, Riggs AD: DNA methylation and gene function. *Science* 1980; 210:604–610.
26. Bird AP: CpG-rich islands and the function of DNA methylation. *Nature* 1986; 321:209–213.
27. Struhl K: Helix-turn-helix, zinc-finger and leucine zipper motifs for eukaryotic transcriptional regulatory proteins. TIBS 1989;14:137–140.
28. Pabo CO, Sauer RT: Protein-DNA recognition. *Ann Rev Biochem* 1984;53:293–321.
29. Pavletich NT, Pabo CO: Zinc finger DNA recognition: Crystal structure of Zif268-DNA complex at 2.1 a. *Science* 1991;252:809–817.
30. McKnight SL: Molecular zipper in gene regulation. *Sci Am* 1991;4:54–64.
31. Cynan WS: Modularity in promoters and enhancers. *Cell* 1989;58:1–4.
32. Ptashne M: Gene regulation by proteins acting nearby and at a distance. *Nature* 1986;322:697–701.

33. Lenardo M, Pierce JW, Baltimore D: Protein-binding sites in Ig gene enhancers determine transcriptional activity and inducibility. *Science* 1987; 236:1573–1577.
34. Smith CW, Patton JG, Nadal-Ginard B: Alternative splicing in the control of gene expression. *Annu Rev Genet* 1989;23:527–577.
35. Lambert S, Yu H, Prchal JT, et al: cDNA sequence for human erythrocyte ankyrin. *Proc Natl Acad Sci USA* 1990;87:1730–1734.
36. Bies R, Phelps SF, Cortez MD, et al: Human and murine dystrophin mRNA transcripts are differentially expressed during skeletal muscle, heart, and brain development. *Nucl Acid Res* (submitted).
37. Vybiral T, Williams JK, Winkelmann JC, et al: Human cardiac and skeletal muscle spectrins: Differential expression and localization. *Cell Motil Cytoskeleton* 1992;21:291–304.
38. Berg DE, Howe M (eds): *Mobile DNA*. Washington, DC: American Society for Microbiology, 1989.
39. Kariya Y, Kato K, Hayashizaki Y, et al: Revision of consensus sequence of human Alu repeats: A review. *Gene* 1987;53:1–10.
40. Ullu E, Tschadi C: Alu sequences are processed 7SL RNA genes. *Nature* 1984; 312:171–172.
41. Nelson DL, Ballabro A, Victoria MF, et al: Alu primed polymerase chain reaction for the regional assignment of 110 yeast artificial chromosome clones from the human X-chromosome: Identification of clones associated with a disease locus. *Proc Natl Acad Sci USA* 1991;88:6157–6161.
42. Donehower LA, Slagle BL, Wilde M, et al: Identification of a conserved sequence in the non-coding regions of many human genes. *Nucleic Acids Res* 1989;17:699–710.
43. Edwards A, Civitello A, Hammond HA, Caskey CT: DNA typing and genetic mapping with trimeric and tetrameric tandum repeats. *Am J Hum Genet* 1991;49:746–756.
44. Little PFR: Globin pseudogenes. *Cell* 1982;28:683–684.

Chapter 3

Techniques of Molecular Biology

The purpose of this chapter is to acquaint both the trainee and the cardiologist with the techniques of molecular biology. As a result of the advancements in molecular biology and its application to clinical cardiac disease, the need to understand this science has become obvious. A variety of methods have been developed to study molecular biological mechanisms and to isolate and characterize genes that cause cardiac disease. The development and application of these molecular techniques to the study of cardiac disease will provide information concerning etiology and pathogenesis. Application of these techniques to inherited cardiac disease will afford us the ability to diagnose these disorders prenatally and postnatally, as well as presymptomatically, and help us in the development of more specific and definitive therapeutic modalities. Chapter 2 should lay the foundation to appreciate the basis for the commonly used techniques in molecular biology. Nevertheless, one property of the nucleic acids is so fundamental and essential to the techniques used, whether they be gel electrophoresis, Southern blotting, or cloning, that it needs to be restated: the complementary base pairing of strands of DNA to each other or of DNA to RNA. The specificity of A═T and C═G binding provides the means to utilize fragments of DNA or RNA as probes to identify, isolate, or characterize other large specific DNA or RNA molecules of interest. The following section describes the commonly used methods available to the molecular biologist.

There are many techniques used in molecular biology; however, some of them are of more universal application than others and are likely to be used routinely even by the medical specialist. It will be helpful for the future to have some understanding of how one isolates DNA or RNA and determines its size and the sequence of its nucleic acids, as well as knowing how to determine in vivo molecular function, as with site-directed mutagenesis in cells or transgenic animals, or how to detect a mutation and determine if it causes a disease. We have selected the techniques of DNA digestion and electrophoresis, Southern and Northern blotting, hybridization, cloning, sequencing, oligonucleotide synthesis, and polymerase chain reaction for discussion. There is already enhanced interest in hereditary diseases of the heart that is likely to continue to accelerate throughout this decade. For that reason we have provided information on reverse genetics (positional cloning), transformation of lymphocytes, restriction fragment length polymorphisms, restriction mapping, chromosomal walking, linkage analysis, allele-specific oligonucleotides, S1 nuclease mapping, and the formation of genomic and complementary DNA (cDNA) libraries.

General Techniques of Recombinant DNA Technology

Isolation of DNA

Essential to molecular genetics is the availability of nucleic acid for study. In the study of human disease, isolation of human DNA is mandatory. Any source may be used for DNA since it is the same in all cells; however, a convenient cell type to use is the leukocyte (1). These cells are easy to obtain, and enough DNA (approximately 200 μg) can be obtained from 10 to 20 ml of blood for genetic analysis or for production of a genomic DNA library. High-molecular-weight DNA may be isolated using detergent treatment, which causes white cell lysis, followed by centrifugation to collect the nuclei. These nuclei are next disrupted using sodium dodecyl sulfate (SDS), with subsequent removal of attached proteins by proteinase K treatment. Extraction of DNA is performed utilizing phenol and chloroform followed by ethanol precipitation. In recent years, DNA isolation has become automated, which yields high-quality DNA. Human DNA may also be recovered from solid organs. The heart is one organ from which DNA may be isolated after homogenization of cardiac tissue obtained via biopsy or from the explanted heart.

Digestion of DNA by Restriction Endonucleases

Restriction endonucleases are enzymes obtained from a variety of bacterial species. These bacterial enzymes are used as protection against foreign DNA, which they achieve by degrading any invading DNA molecules. Each enzyme recognizes a specific sequence of three to eight nucleotides in foreign DNA (Table 3.1). There are two major types of restriction endonucleases. Type I enzymes recognize specific nucleotide sequences but their cleavage sites are nonspecific (2) and therefore are, in general, not useful for DNA technology. Type II enzymes, in contrast, recognize a particular target sequence in a duplex DNA molecule, thereby breaking polynucleotide chains within that sequence to create discrete DNA fragments (3,4). These Type II restriction enzymes will cut any length of DNA double helix at their recognition sites into a series of fragments, known as restriction fragments, each of which often differ in size depending on the distribution of these "cut" sites along the linear DNA (Fig. 3.1).

Table 3.1 Some Restriction Enzymes and Their Recognition Sites

Microorganism	Abbreviation	Sequence of Recognition and Cleavage Site (5′ → 3′) (3′ → 5′)
Escherichia coli RY13	Eco R1	GAATTC CTTAAG
Haemophilus influenza Rd	HindIII	AAGCTT TTCGAA
Providencia stuartii 164	Pat I	CTGCAG GACGTC
Bacillus amyloliquefaciens H	Bam H I	GGATCC CCTAGG
Haemophilus aegyptus	Hae II	PuGCGCPy PyCGCGPu
Streptomyces albus G	Sal I	GTCGAC CAGCTG
Haemophilus influenzae Rd	HindII	GTPyPuAC CAPuPyTG

The DNA nucleotide sequences recognized by restriction enzymes are typically "palindromic"; that is, the nucleotide sequences of the two strands are symmetrical on either side of the cut site in the recognized region. The two strands of DNA are cut at or near the recognition sequence, often with a staggered cleavage, creating cohesive ends both of which are short and single stranded. These cohesive or "sticky" ends (a DNA molecule with single-stranded ends that show complementarity and allow the molecule to join end to end with introduced fragments) can form complementary base pairs with any end produced by the same or other enzymes that cut at the identical sequence site. Enzymes that create sticky ends include *Eco*RI and *Hind*III, as well as the majority of other restriction enzymes (Fig. 3.2). The other type of restriction endonucleases cut DNA at their recognition sites but create blunt-ended fragments (Fig. 3.3) that are base paired to their ends (i.e., no nucleotides overhanging the ends). These fragments have no tendency to stick together. *Hind*II and *Hpa*I are examples of blunt end–generating enzymes. Since there are over 100 commercially available enzymes, a large number of different fragments can be generated from specific DNA stretches. These commercially available enzymes are sold at known concentrations and their activity is defined as one unit being equal to the amount of enzyme required to cut 1 μg of DNA in 1 hour. Each enzyme works optimally at certain known temperatures using specific buffer mixtures. Selection of the appropriate endonucleases depends on the size of the fragments desired; 3–base pair (bp) cutters give rise to many fragments whereas 8-bp cutters give rise to fewer fragments. It also depends on the number of sites present, so a process of trial and error may be required to find

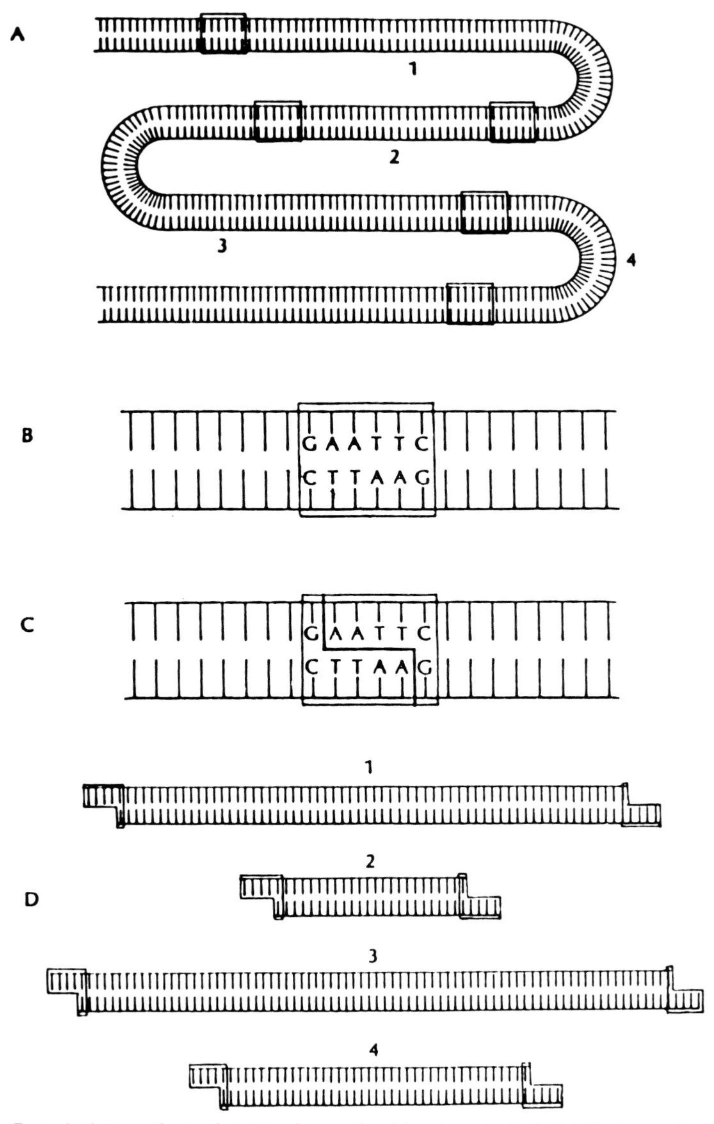

Figure 3.1 Restriction endonucleases cleave double-stranded DNA (A) at specific restriction sites (different for each enzyme). Typically, the sites are short sequences of three to eight bases. The sequence on one DNA strand is the reverse of the other (B). The cleavage (C) produces fragments of various sizes (D). A mutation that involves the restriction site would alter the number of fragments as well as their size.

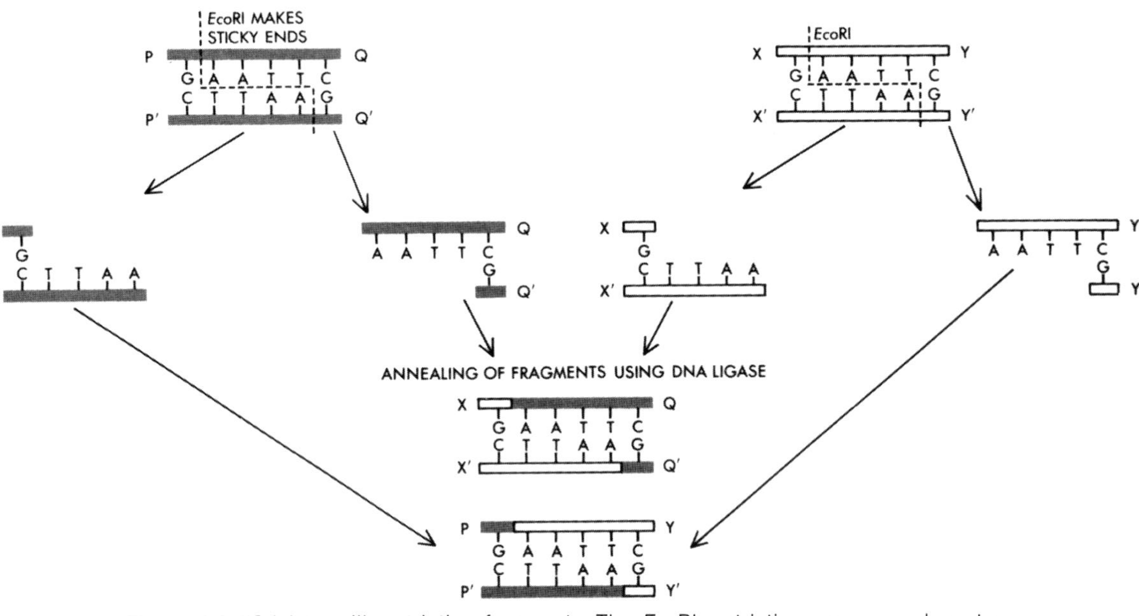

Figure 3.2 "Sticky end" restriction fragments. The *Eco*RI restriction enzyme makes staggered, symmetrical cuts in DNA away from the center of its recognition site, leaving cohesive or "sticky" ends. A sticky end produced by *Eco*RI digestion can anneal to any other sticky end produced by *Eco*RI cleavage.

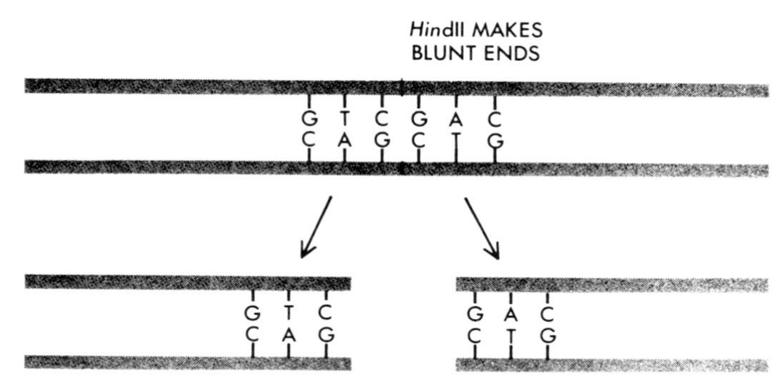

Figure 3.3 An example of a restriction endonuclease (*Hin*dII) that digests with the formation of blunt ends.

the most appropriate restriction endonuclease. Since the enzyme cut is sequence specific, some DNA segments may not contain a cut site while other segments do. This is one simple way to distinguish between two pieces of DNA that are different.

Separation of DNA Fragments by Gel Electrophoresis

After the DNA is cleaved by the restriction endonuclease, the digested DNA is then loaded into a well of an agarose or polyacrylamide gel and subjected to electrophoresis (Fig. 3.4). Each individual nucleotide has a net negative charge that forms the basis for separation of DNA fragments in an electrical current. The DNA fragments are negatively charged and therefore migrate toward the anode according to their size, with the larger fragments migrating slowest and remaining nearest the wells at the top of the gel. Electrophoresis is the technique routinely used to separate DNA fragments. Electrophoresis is performed for sufficient time to separate the fragments, and the gel is then stained with ethidium bromide (which intercalates between bases and fluoresces under ultraviolet illumination) and photographs of the DNA fragments are obtained. The fragments of DNA may appear as a smear in the lane if there is a large amount of various-sized DNA within each lane. If only a few discrete fragments are separated, specific identifiable bands will be seen on the gel. To determine the size of the fragments of the unknown DNA, standard DNA fragments of known size are concomitantly electrophoresed for comparison (Fig. 3.5). Electrophoresis through agarose will separate double-stranded DNA fragments varying from 70 bp to 100,000 bp, and polyacrylamide is used to separate DNA fragments from 6 to 1,000 bp. Utilizing these gels, one can detect bands of as little as 1 ng and a difference between fragments of 0.5% of their size. Polyacrylamide can separate DNA fragments with a single base pair size difference.

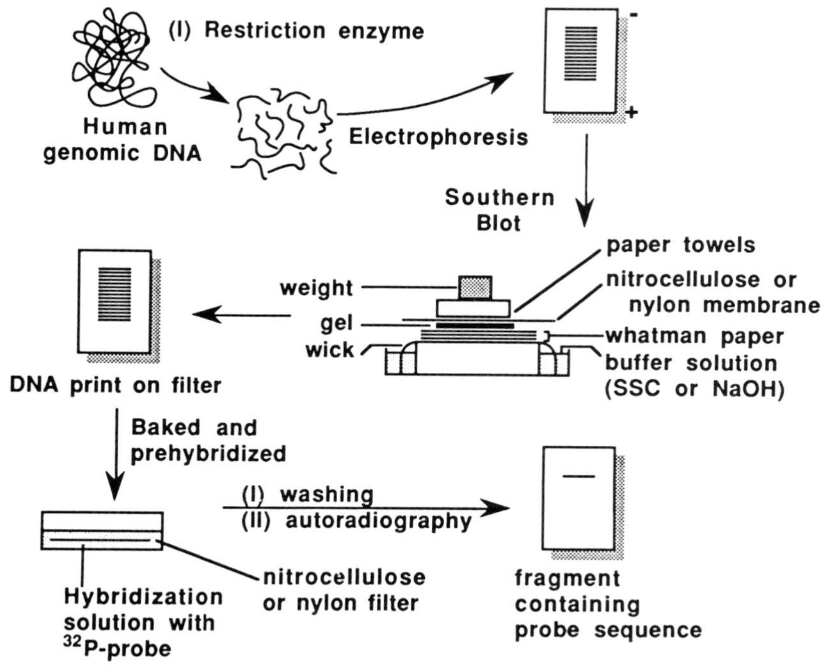

Figure 3.4 Southern blot analysis of genomic restriction enzyme DNA fragments. Initially the genomic DNA is digested with restriction enzyme, loaded into the wells of an agarose gel, and then electrophoresed. The DNA is then transferred from the gel to a nylon or nitrocellulose filter by capillary flow. This filter is later hybridized to a complementary probe and autoradiographed.

Southern Blotting

Southern Blotting, named after its inventor, E. M. Southern (5), utilizes several techniques, including restriction digestion of the DNA and separation of the DNA fragments by electrophoresis as discussed earlier. Southern discovered that, following separation of DNA by electrophoresis on agarose or polyacrylamide gel, the DNA can be transferred in a buffer from the gel through capillary action to a nitrocellulose filter (Fig. 3.6). The nitrocellulose filter provides a hard copy for the DNA separation accomplished in the gel system. In essence, one fills a flat tray with buffer, which is partially covered with a glass or plastic cover upon which is laid thick filter paper, a portion of which extends around the cover into the buffer to provide a wick. Usually another piece of filter paper overlaps this, upon which is laid the gel. The gel is overlaid by a nitrocellulose or nylon filter on top of which is placed a paper blotter (paper towels) that provides the capillary action to pull the buffer through the gel to the filter. The flow of buffer from the gel to the filter is perpendicular to the direction of electrophoresis. This flow causes the DNA fragments to be carried out of the gel onto the filter, where they bind and give a replica

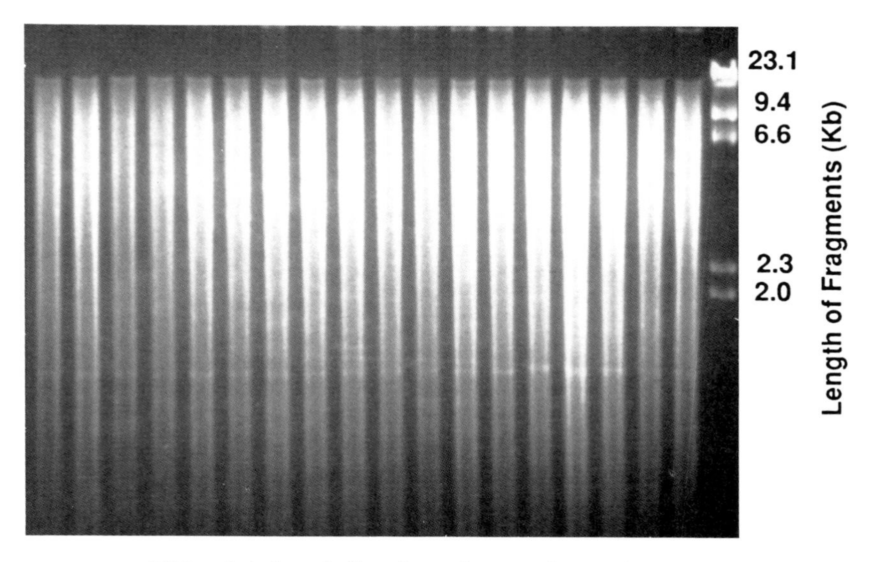

Figure 3.5 Electrophoresis of DNA following digestion with restriction endonucleases, with each vertical lane corresponding to a separate DNA sample. The staining is with ethidium bromide, which stains all of the bands, and therefore produces more of a smear.

(or "print") of the DNA fragments as they were separated on the gel (Fig. 3.7). The reason for the DNA binding to the filter is not known. After the transfer, the DNA is fixed to the filter by baking or crosslinking with ultraviolet light and remains essentially permanently attached to the nylon or nitrocellulose. The DNA is now ready for hybridization to a labeled DNA probe. Nylon is preferred today over nitrocellulose since it lasts much longer and is much less brittle. In the process of hybridization, the nylon filter containing the DNA and the labeled DNA probe are placed in a hybridization solution, usually in polyethylene bags that are carefully heat sealed. If the labeled probe is complementary to the DNA of interest, it will hybridize (anneal) under the proper conditions; washing is then performed to remove nonspecific binding, followed by autoradiography to detect the labeled probe–DNA complex. The hydrogen binding between the DNA fixed to the nylon and that of the labeled DNA probe can be broken with detergent (referred to as "stripping"), so the sample of DNA fixed to the filter can be used many times to hybridize to new probes.

Southern analysis is a very important and common procedure in molecular biology and has in common a rational basis shared by two other techniques, Northern and Western blotting. Thus, the steps for Southern blotting are summarized to emphasize the rationale and the time required to complete the procedure:

1. Preparation of the sample, including restriction enzyme digestion (4 to 24 hours).

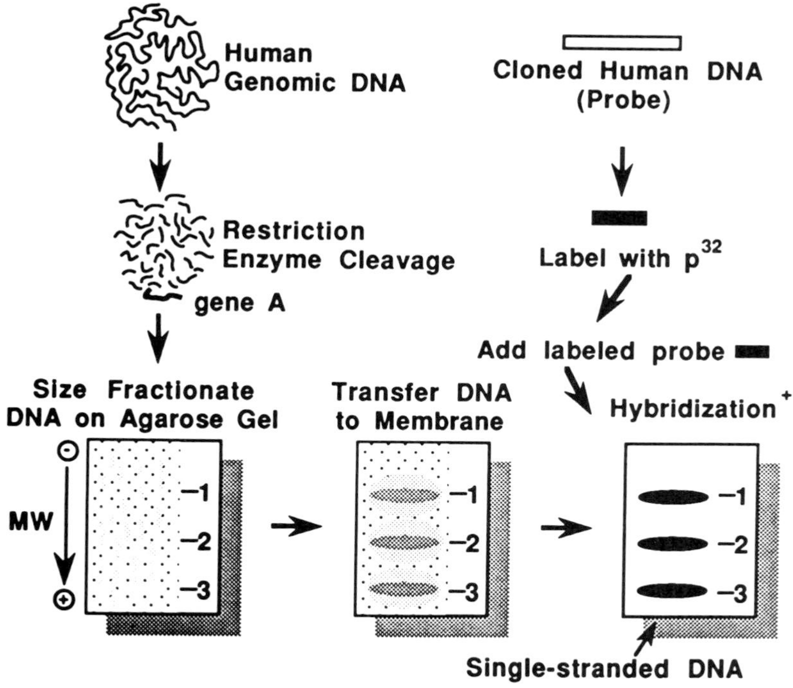

Figure 3.6 Illustration of the Southern blotting technique. The DNA is cleaved with an appropriately selected restriction endonuclease. The digested fragments are separated by electrophoresis on agarose gel, and the fragments of gene A are located at positions 1, 2, and 3 but cannot be seen against the background of many other randomly occurring DNA fragments. The DNa is denatured and transferred to a membrane in an identical pattern to what it was on the agarose gel. It is difficult to manipulate anything on a soft gel or to remove it. Once transferred to the membrane (filter), a solid support system, the DNA is much easier to handle. A DNA probe (cDNA) that has been labeled with ^{32}P is hybridized to its complementary DNA and visualized after exposure of the nylon membrane to an autoradiograph. The method of transfer of the DNA from the gel to the membrane, developed by Southern, was a major innovation (see Fig. 3.7).

2. Separation of the DNA fragments by electrophoresis (3 to 24 hours).
3. Denaturation of the double-stranded DNA into single-stranded DNA by NaOH (1 to 2 hours).
4. Transfer of DNA fragments from the gel to the nylon filter (12 to 48 hours).
5. Fixation of the DNA to the filter (less than 1 hour).
6. Hybridization (8 to 12 hours).
7. Washing, air drying, and exposing the filter to x-ray film (12 hours to 5 days).

Completion of the procedure usually requires 7 to 10 days, although this is dependent on the DNA analyzed and the specific activity of the probe, which may allow for results within 3 days.

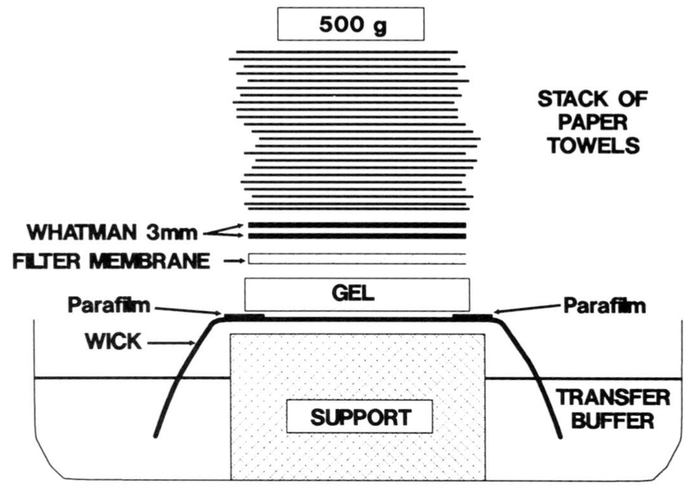

Figure 3.7 Southern transfer apparatus: 1) tray filled with 20× SSC; 2) glass plate (supported on two sides of the tray); 3) wick of three sheets of Whatman 3MM paper; 4) gel; 5) Parafilm around all sides of the gel; 6) filter; 7) three sheets of Whatman 3MM paper; 8) paper towels; 9) glass plate; and 10) weight. The paper towels provide the necessary absorbent to initiate the capillary action required to transport the DNA from the gel to the filter. The filter is then baked or exposed to ultraviolet light to permanently link the denatured DNA covalently to the filter.

Previously we showed how staining DNA fragments with ethidium bromide subsequent to their separation by electrophoresis may create essentially a smear. In the Southern blot technique only the few discrete fragments that bind to the probe are seen when the filter with a DNA smear is exposed to x-ray film. Since the probe binds only to DNA fragments that exhibit complementarity, only these duplex fragments will be visualized on the autoradiograph (Fig. 3.8).

Northern Blotting

This blotting method, the RNA equivalent of Southern blotting, was described by Alwine and colleagues (6) in 1977. The RNA is fractionated on an agarose gel and transferred by blotting to nitrocellulose or nylon filters. A radiolabeled DNA probe (usually cloned cDNA or genomic DNA) allows detection of the corresponding RNA sequence bound to the filter (Fig. 3.9). If RNA is extracted from the nuclei, the sizes of the precursors can be determined. An estimate of the abundance of mRNA can be made and the response to hormonal or metabolic stimuli can be followed. Care must be taken when performing this procedure since ribonucleases (RNAses; enzymes that degrade RNA) are ubiquitous and difficult to erad-

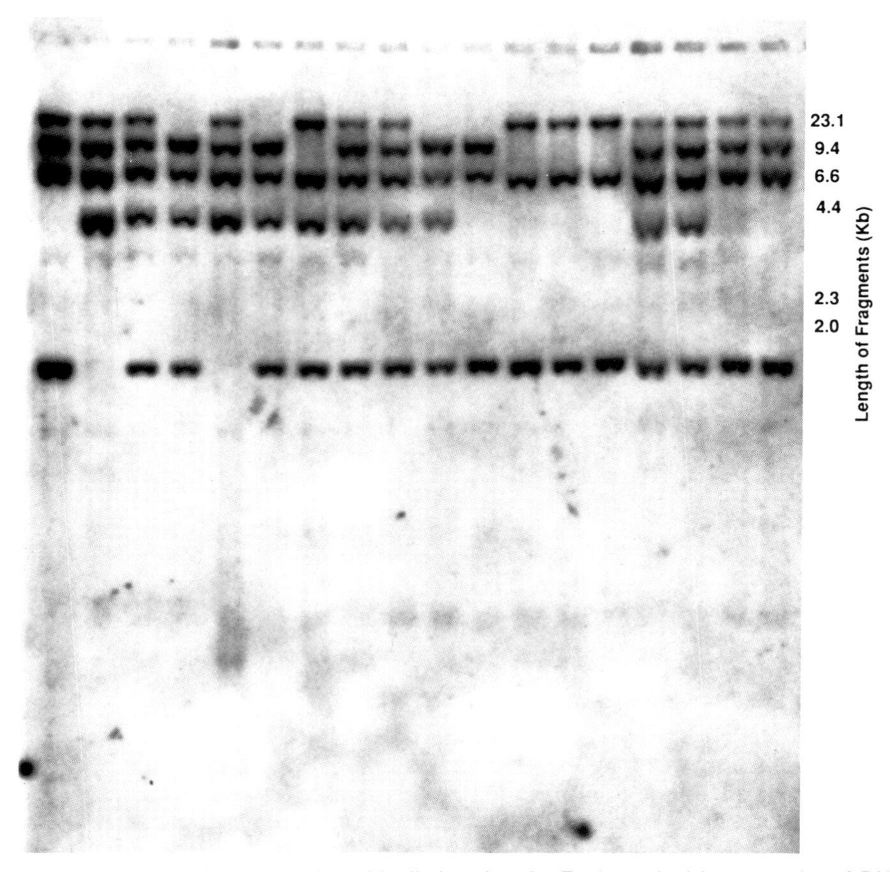

Figure 3.8 A typical Southern blot with distinct bands. Each vertical lane consist of DNA from a separate individual. All of the DNA samples were digested with the same restriction endonuclease. Following separation on electrophoresis and transfer to a nylon membrane, hybridization was performed with the selected radioactive probe, and thus only those fragments complementary to the probe are visualized. This is in sharp distinction to staining with ethidium bromide (shown in Fig. 3.27). This is an analysis of a family with hypertrophic cardiomyopathy, and the different patterns reflect restriction fragment length polymorphisms (RFLPs) characteristic of the marker locus, which is linked to the disease locus.

icate. New glassware or glassware used only for RNA work is generally required.

Isolation of RNA

RNA extraction is best performed from blood or tissue frozen freshly in liquid nitrogen or dry ice, since RNA has a tendency to degrade quite rapidly. The typical mammalian cell contains about 10^{-5} μg of RNA, 80% to 85% of which is ribosomal (28S, 18S, and 5S) and 10% to 15% of which is made up of a variety of low-molecular-weight species that include transfer RNAs (tRNAs), small nuclear RNAs, and others. All of these RNAs

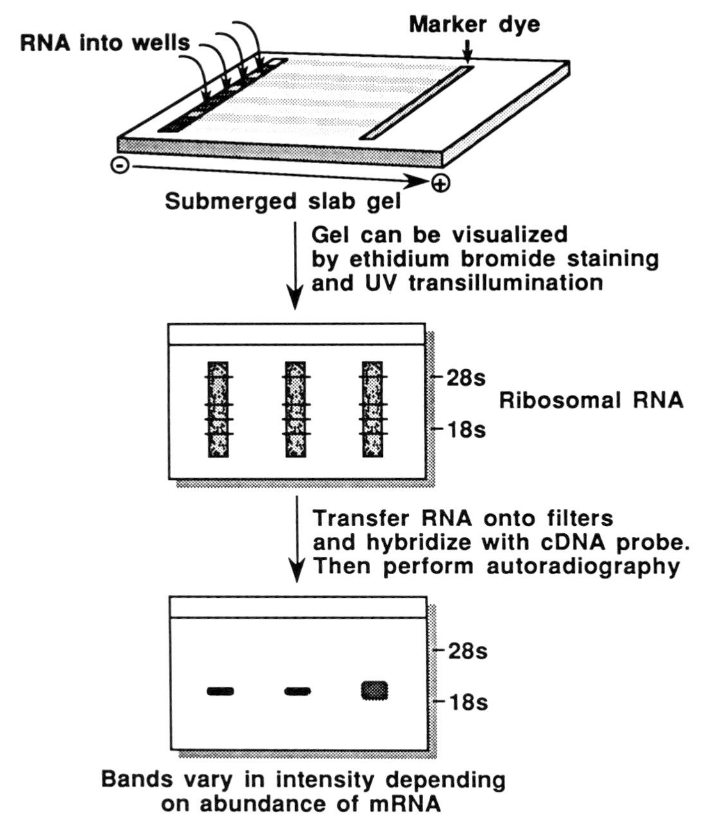

Figure 3.9 A flow diagram of Northern blotting (which is virtually identical to Southern blotting except the sample is RNA).

are of defined size and sequence, and their isolation in virtually pure form is possible utilizing gel electrophoresis, density gradient centrifugation, or ion exchange chromatography. By contrast, mRNA, which makes up 1% to 5% of the total cellular RNA, is heterogeneous both in size (ranging from several hundred base pairs to several kilobase pairs in length) and in sequence. However, essentially all mammalian mRNAs carry at their 3′ ends a poly(A) tail that is generally long enough to allow mRNA to be purified via utilization of affinity chromatography on oligo(dT)-cellulose columns. The resulting heterogeneous population of molecules collectively encodes all the polypeptides synthesized by the cells. It is necessary to limit the RNAse activity during the early stages of RNA extraction as well as to avoid accidental introduction of RNAses via glassware and solutions. Proteins dissolve readily in solutions of potent denaturing agents such as guanidine HCl and guanidinium thiocyanate (7). Cellular structures are destroyed and nucleoproteins dissociate from nucleic acids as protein secondary structure is lost. RNAses are inactive when exposed to guanidinium thiocyanate (4 mol/liter) or β-mercaptoethanol (8), and these are commonly used. Total cellular RNA isolation has been performed in a number of ways in the past.

First, the procedure initially described by Favaloro et al. (9) for isolation of RNA from mammalian cells in tissue culture is a tried and true method. The growing cells are first lysed, then proteins are removed by phenol: chloroform extraction and centrifugation, and finally RNA is recovered by precipitation. More recently, a number of rapid isolation methods of RNA from cultured mammalian cells have been developed. The two major methods are those dependent on differential extraction of RNA by organic solvents such as acidic phenol (10) and those utilizing differential precipitations to separate high-molecular-weight RNA from other nucleic acid (11). For tissue extraction of RNA, strong detergents are generally utilized.

The most common method is the guanidinium thiocyanate–cesium chloride centrifugation method (12,13). This technique begins by tissue homogenization in guanidinium thiocyanate, followed by addition of cesium chloride to the homogenate. Centrifugation is next performed in order to take advantage of the fact that the molecular densities of RNA and cesium chloride are much greater than those of other cellular macromolecules, thereby allowing separation. During this centrifugation, the RNA forms a pellet on the bottom of the tube while the DNA and protein float upward in a cesium chloride solution. After discarding the supernatant, the RNA pellet is dissolved in a buffer mix, then extracted with chloroform and butanol. Precipitation of this RNA is then performed using sodium acetate and ethanol at −20°C. More recently, the RNAzol method (14) has been used to extract RNA more rapidly from tissue.

Hybridization and Development of a Probe

Almost all of the procedures used in molecular biology exploit DNA-DNA or RNA-DNA hybridization and require a labeled DNA probe. To isolate a specific fragment of DNA or RNA, whether as part of a Northern or Southern blotting procedure, requires having a DNA probe that is labeled with a tag that is easy to detect, such as a radioactive compound or a fluorescent dye, or is coupled to an enzyme. The probe can be used to isolate its complementary DNA or RNA through a process referred to as molecular hybridization. The formation of a double helix from two complementary strands (annealing or hybridization) was first described by Marmur and Doty in 1960 (15). In the nucleus the double helix has 100% complementarity—that is, A pairs with T, and C pairs with G (16) or, in the case of RNA, A pairs with U—but hybridization can occur even if some of the bases of the duplex do not bond (e.g., A opposite G or C opposite T). In the latter pairing molecular bonding cannot occur, but hybridization occurs as long as most of the other base pair alignments are appropriate for molecular bonding. However, the stability of the duplex formed is directly related to the extent of the complementarity. There are various forms of hybridization in use: solution hybridization, filter hybridization (Southern and Northern), and in situ hybridization (e.g., biopsy material fixed to a glass slide for microscopic analysis). Any fragment of nucleic acid of known size can theoretically be labeled as a probe, but usually single-stranded DNA rather than RNA is used. RNA is less stable, being easily degraded by nucleases; in addition, with RNA it is difficult to exclude impurities, which give background noise, and yields are very low since there is only one means of labeling RNA: ^{32}P, with T4 polynucleotide kinase. However, occasionally mRNA is the only available probe.

The probe (labeled DNA), under appropriate conditions, is used to identify and isolate another DNA molecule, which is usually larger. The probe hybridizes if there are complementary bases on the DNA of interest. The probe may be synthesized DNA, referred to as an oligonucleotide (described later), or it may be a natural fragment of DNA that had been previously cloned.

The methods of labeling DNA include nick translation, oligohexamer or random hexamer labeling, and kinase labeling.

Nick Translation (17,18)

DNAse I introduces nicks at widely separated sites in DNA, exposing a free 3′ hydroxyl group, which allows DNA polymerase I of *Escherichia coli* to incorporate nucleotides successively. Concomitant hydrolysis of the 5′ terminus by the 5′ → 3′ exonucleolytic activity of polymerase I releases 5′ mononucleotides. If the four deoxynucleoside triphosphates are radiolabeled with ^{32}P, the reaction progressively incorporates the label into a duplex that is unchanged except for translation of a nick along the molecule. Kits are commercially available from a variety of manufacturers.

Oligohexamer or Random Hexamer Labeling (19)

This labeling scheme produces probes of very high specific activity (i.e. 10^9 cpms/μg of DNA. It does so by denaturing the DNA and then combining random hexadeoxynucleotide primers together with the Klenow fragment of DNA polymerase and all four nucleotide triphosphates, one or more of which will be radiolabeled. The Klenow fragment, the larger of the two fragments produced when DNA polymerase I is cleaved by subtilisin, retains it 5′ → 3′ polymerase activity while losing the 5′ → 3′ exonuclease activity. This enzyme produces a radiolabeled DNA molecule complementary to the nonradioactive, denatured DNA. Care must be taken when using the Klenow fragment, because it is somewhat finicky and requires gentle handling. Kits are available from a variety of commercial manufactures.

Kinase Labeling (20,21)

This method involves labeling the 5′-end of DNA using T4 polynucleotide kinase after the 5′ terminus is dephosphorylated. This method, also known as end labeling, is commonly used to label short oligonucleotides. The number of counts per minute obtained by this labeling scheme is significantly lower than that described, for instance, for random hexamer labeling. The labeled probe is incubated with the nylon filter containing the DNA of interest in an appropriate hybridization solution. Usually the filter, probe, and solution are placed in a polyethylene bag that is carefully heat sealed. The Church and Gilbert (22) method is usually employed and consists of a simple hybridization solution containing sodium phosphate, EDTA, and SDS. If a labeled probe that is complementary to the DNA of interest is used, it will anneal (i.e., hybridize) under the proper conditions. Once hybridization is completed, the excess hybridization solution and radioactive label are washed off with a solution containing the detergent SDS. The higher the stringency conditions used for hybridization and washing, the lower the background typically seen on autoradiography. However, this lower background is obtained at the risk of decreasing the specific signal provided by the labeled probe. Higher stringency is obtained by using higher temperatures and lower salt conditions.

Histological In Situ Hybridization

This approach has a major advantage over solution hybridization for the quantitation of copy numbers of message per cell. In addition, hybrid molecules (i.e., DNA-DNA, DNA-RNA, RNA-RNA) formed between the nucleic acids immobilized in cytohistological preparations can be studied to obtain information regarding gene expression within a heterogeneous cell population. RNA species present in as low an amount as 0.01% can be detected using this method. In addition, while examination of stage specificity of mRNA populations in early embryos is difficult by Northern analysis because insufficient quantities of RNA are isolated, in situ techniques can be used to evaluate individual cells for their information content. Significantly better resolution may be obtained by this method for detection of individual RNAs in any cell type at any developmental stage.

In situ hybridization has been used in the diagnosis of myocarditis from endomyocardial biopsy specimens by Bowles et al. (23), allowing diagnosis of coxsackie B myocarditis. In addition to being a diagnostic aid, in situ hybridization also demonstrated virus-specific RNA late in the course of disease, suggesting that viral replication continues throughout the various stages of disease and demonstrating that the viral genome remains in tissue for long periods of time.

DNA Cloning and Gene Libraries

DNA fragments from any source can be amplified more than 10^6-fold by inserting them into a plasmid or bacteriophage, referred to as a vector, and then growing them in a suitable host, such as bacterial or yeast cells (Fig. 3.10). Plasmids are small circular molecules of double-stranded DNA naturally occurring in bacteria and yeast, where they independently replicate as the whole cell proliferates. Despite generally accounting for only a small percentage of the total DNA in the whole cell, they often carry vital genes. Because of its small size, plasmid DNA can easily be separated from host cell DNA and purified. These purified DNA molecules may be used as cloning vectors after being cut by the restriction enzymes and then ligated to the DNA fragment to be cloned. The resultant hybrid plasmid-DNA molecules are then reintroduced into bacteria that have been transiently made permeable to macromolecules, but only a portion of the treated cells actually take up the plasmid. The cells will survive in the presence of the antibiotic(s) whose resistance genes are encoded by the plasmid. These bacteria divide with concomitant plasmid replication to produce large numbers of the copies of the original DNA fragment. The hybrid plasmid-DNA molecules are then purified and the copies of the original DNA fragment excised by restriction enzyme digestion. The cloning process, therefore, may produce millions of different bacterial or yeast colonies, each harboring a plasmid with a different inserted DNA sequence. The rare colony whose plasmid contains the genomic DNA fragment of interest must then be selected (Fig. 3.11) and allowed to proliferate to form a large cell population or clone. Identification of the colony of interest involves use of radioactive nucleic acid probes whose nucleic sequence is complementary to the desired cloned DNA (24,25).

Gene libraries are large collections of individual DNA fragments growing in a suitable host such as *E. coli*. These may be either a genomic library (which is made up of fragments of nuclear DNA), a chromosomal library (made up of fragments derived from a specific chromosome), or a cDNA library having expressed sequences derived from the total mRNA population of the cell. After growing a library, which may contain more than 10^6 recombinant DNA molecules, in the appropriate host on agarose plates, the library is transferred to a nitrocellulose or nylon membrane by the technique of replica plating (Fig. 3.11). In this method, after transfer of the DNA from the colonies to the filter, the DNA is denatured by alkali. The filter is then baked in an oven, with the DNA remaining tightly bound to the membrane and representing an exact copy of the DNA sequences present on the original agarose plates. The membrane is then

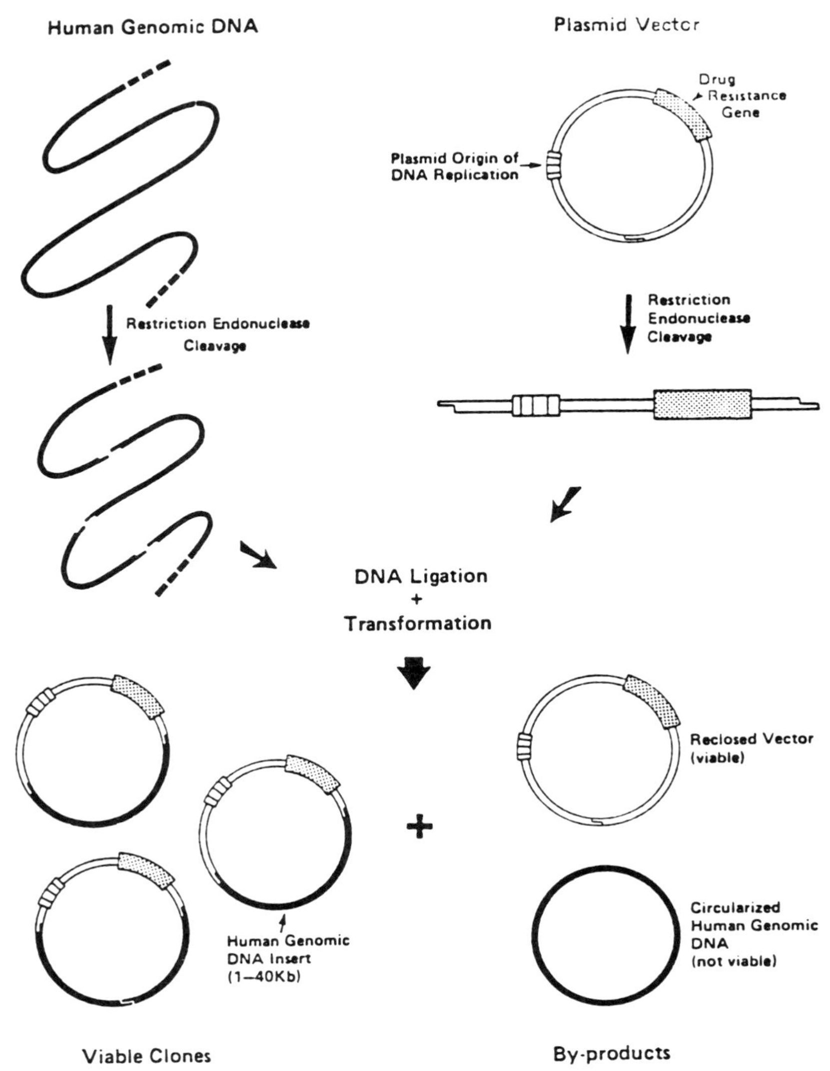

Figure 3.10 The cloning of a DNA fragment utilizing a plasmid as the vector. Initially the human genomic DNA is cleaved with a restriction enzyme and then inserted into the plasmid vector that contains a means of replication and a selectable marker, such as a drug resistance gene. Various by-products of the procedure (i.e., closed vector and circularized human DNA) are also represented.

hybridized with the labeled probe and autoradiographed. The clones that hybridize with the probe are seen as darkened replicas of the clone on the autoradiogram, which corresponds to a spot on the agarose plate. These positive clones may be picked out and analyzed by Southern blotting. This method of colony hybridization, first described in 1975 by Grunstein and Hogness (24), was the first technique that allowed for isolation

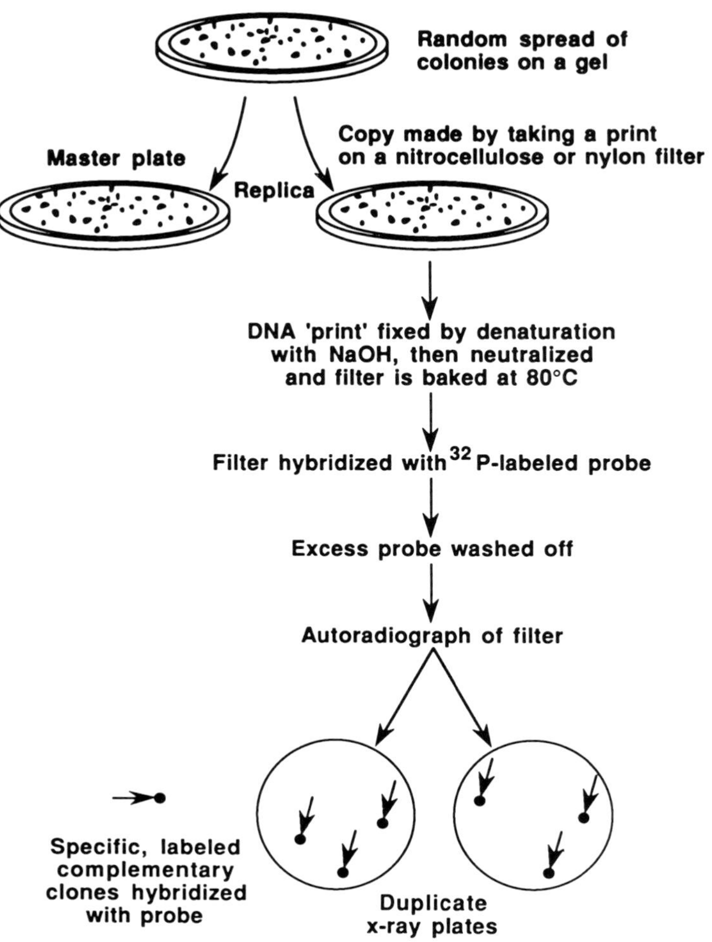

Figure 3.11 Flow diagram for identification of recombinants following cloning by the hybridization method. The bacteria from Figure 3.10 that one hopes contain the recombinant DNA of the vector are grown in colonies on agarose contained in routine petri dishes. Once the colonies are visible, one can replicate the colonies in the petri dish using a nitrocellulose filter or nylon membrane. A circular piece of nylon the size of the petri dish is placed over the colonies and three asymmetrical marks are placed on the membrane and dish. The membrane is lifted from the plate and then removed. The filter bearing the colonies is treated with alcohol so that the colonies are lysed and their DNA released and denatured. The filter is then treated with proteinase to remove the protein. The denatured DNA is fixed firmly to the filter. The ^{32}P-labeled DNA, is incubated with the filter so that it will hybridize to its complementary DNA. Following this, the filter is washed to remove any excess label and is then exposed to x-ray film (autoradiography). After several hours or several days of exposure, the film is developed, and should show spots that mirror the sites of those colonies on the original plate that have the human insert. By referring the pattern back to the master plate (matching the markers), one can see which colonies do indeed have the insert, and these can be subplated and serve as a renewable source of colonies that have the human insert. A similar procedure can be followed if one uses bacteriophage that form plaques rather than colonies.

of cloned DNAs containing a specific gene and has helped to revolutionize molecular genetics. Another technique that is increasingly used today is color coding of the colonies through an appropriate reporter gene attached to the gene of interest. All of the colonies that have incorporated the DNA to be cloned will also have the accompanying gene, which implies that there will be a specific color to the colonies for each hybridization.

The identification of recombinants can be performed using a variety of different probes. These include radiolabeled oligonucleotide probes, cDNA probes, and immunological methods for detection and expression.

DNA Sequencing

Two classical methods of DNA sequencing are in common use, with an analysis potential of up to 1,000 bp daily.

Maxam-Gilbert Method (26)

This chemical method involves isolation of a fragment of DNA labeled at one end with ^{32}P that is then subjected to a set of four partial, but base-specific, cleavages that produce a series of subfragments. These are separated by size and polyacrylamide-urea gels at high voltage, and the labeled fragments are then detected by autoradiography. The base sequence can be read off a sequencing gel autoradiograph from the ladder of each of the base-specific tracks starting from the bottom of the gel (Fig. 3.12). This is presently the method of choice, and automated sequencers are now available.

Sanger-Coulsen Method (27)

This method involves the cloning of the DNA fragments into the single-stranded filament virus M13. Initiation of synthesis of a copy of the inserted fragment whose sequence is desired occurs via a short DNA primer. This synthesis is interrupted by four labeled deoxynucleotides that terminate growth of the chain at any point at which the natural deoxynucleotide should be introduced. The resulting set of products are analyzed in the same gel system as above (Fig. 3.13). Either ^{32}P or ^{35}S can be used.

Synthesis of Polynucleotides

Two types of synthesis are currently in use: 1) the synthesis of short stretches of nucleotides up to 50 bp long (called oligonucleotides), and 2) the synthesis of DNA with sequence complementarity to mRNA sequences—the so-called cDNA or complementary DNA.

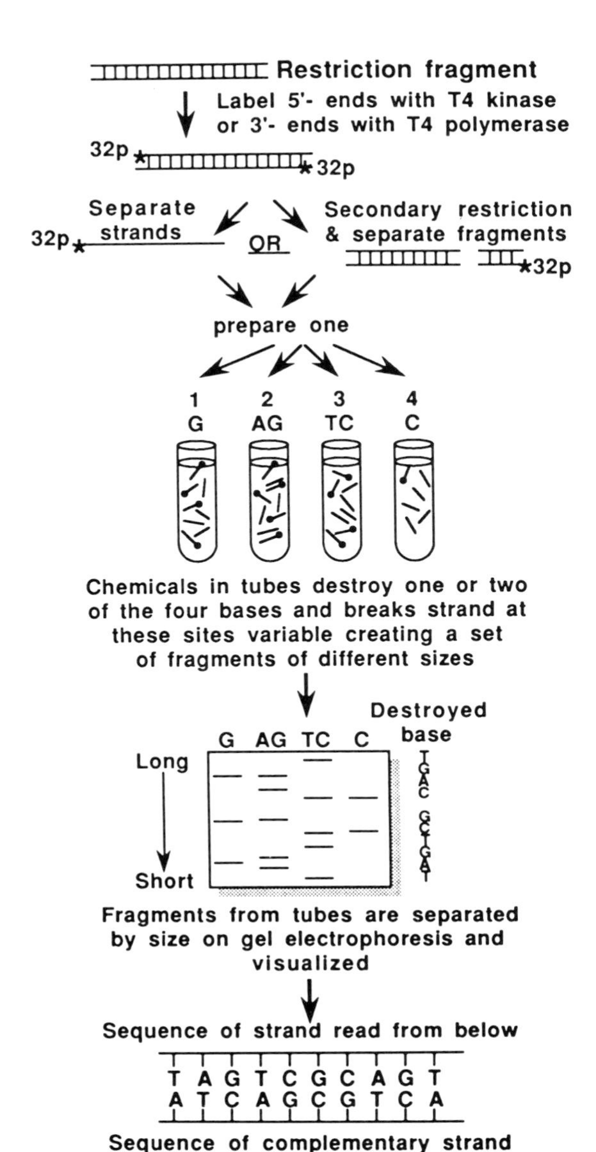

Figure 3.12 DNA sequence determination. Outline of the steps involved in the chemical cleavage procedure of Maxam and Gilbert using terminally labeled single- or double-stranded DNA fragments.

Oligonucleotides

These linear sequences of nucleotides, usually prepared from an amino acid sequence, can be used to initiate DNA synthesis on fractions of mRNA to make a complementary copy (28). More recently, oligonucleotides for polymerase chain reaction (PCR) or allele-specific oligonucleotides (ASOs) have been prepared from known DNA sequences. If the oligonucleotide (antisense) sequence codes for a unique stretch of the peptide of interest, then it may hybridize preferentially with its mRNA to prime the synthesis of cDNA. Another use for oligonucleotides is as probes in

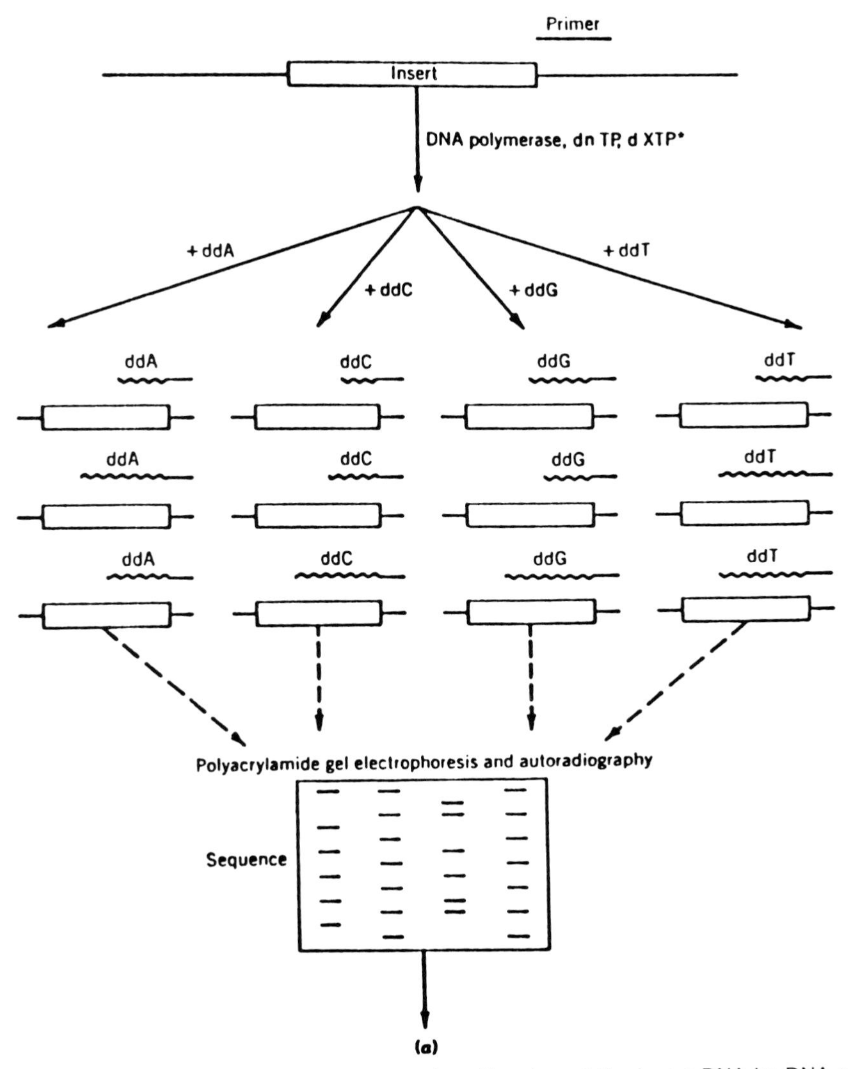

Figure 3.13 (A) Dideoxy reaction for sequencing. Copying of the insert DNA by DNA polymerase is inhibited at specific sites in the presence of dideoxynucleotides. (*Continued*)

the screening of cDNA gene libraries in order to identify clones carrying complementary sequences. Oligonucleotides can be synthesized in cycles, each of which adds one or two bases to the growing chain, with subsequent ligation of these nucleotides. Complex mixtures of oligonucleotides may be created by addition of nucleotides at certain points in the cycle. Thus, synthesis of a pool of oligonucleotides that represent all possible triplicate sequences coding for the amino acids at a particular peptide may be accomplished. This oligonucleotide mixture can then be used as probes to identify DNA sequences coding for the peptide in a

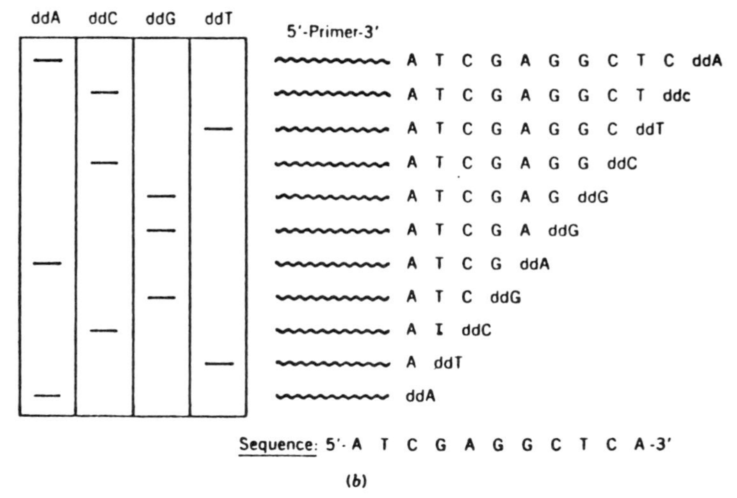

Figure 3.13 (B) The different fragments obtained are separated according to size by electrophoresis. Sequence can then be determined.

cDNA library, thus allowing for isolation and identification of the gene. Oligonucleotides are very versatile. For instance, they may be used for primers in PCR, as probes for Southern analysis, or as ASOs.

Complementary DNA

With a short, single-stranded primer, DNA sequences complementary to RNA can be synthesized (29). This primer can be a stretch of oligothymidines, which will bind to the poly(A) tail of mRNA, or a mixture of random sequence oligonucleotides complementary to multiple segments along the RNA sequence. Reverse transcriptase will copy the mRNA into a complementary strand of DNA, with subsequent removal of the mRNA utilizing aldehyde treatment. The second strand of DNA is then synthesized on the first strand by a Klenow fragment. The resulting double-stranded cDNA can then be trimmed with S1 nuclease to remove the hairpin loop that forms, thus creating a blunt-ended molecule that can be ligated into a plasmid or a phage vector for cloning.

Polymerase Chain Reaction

This technique, devised by Mullis et al. (30), allows for the specific amplification of discrete DNA fragments. This amplification results in easier detection of nucleic acid fragments initially present in very small (picogram) quantities. Significant reduction in time and labor requirements necessary for the analysis and production of desired DNA fragments occurs since the need for subcloning and plasmid amplification is negated. Isolation and purification of nucleic acid fragments is enhanced because the amplified sequences become the most prominent species in the sample. Existing DNA sequencing methods can be coupled to the PCR technique, directly eliminating the need for cloning and purification of the nucleic acid fragments to be sequenced.

Amplification of DNA sequences by the PCR method mimics the natural DNA replication process of doubling the number of molecules after each cycle (Fig. 3.14). The "cycle" here is the repetition of a set of three successive steps performed under different, but controlled, temperatures. The amplified product of interest ("short product") begins to accumulate after as few as three cycles. The "short product" (31,32) is the region between the 5′ ends of the extension primers (synthetic oligonucleotides that anneal to the sites flanking the region to be amplified), which contains discrete ends corresponding to the sequence of these primers. As the cycle number increases, the "short product" become the predominant template to which the extension primers anneal. Theoretically, the amount of amplified product should double after each cycle, leading to exponential accumulation. Because of enzyme kinetics, however, the amount is actually somewhat lower.

As the number of cycles proceeds, other products also form. The "long product," derived in each cycle from the template molecules, has variable 3′ ends. The quantity of product created increases arithmetically throughout the amplification process since the quantity of the original template remains constant. At the end of the amplification process, the short product is typically overwhelmingly more abundant.

To simplify one's understanding of the PCR process, the following points should be remembered. In order to successfully perform PCR, the sequence of the region of interest must be known. Oligonucleotide primers are prepared for specific portions of the sequence at the 5 ′ end and the 3′ end in order to amplify a piece of DNA of interest that lies between these primers. The primers must be sequence specific at both ends or there must be a sequence-specific primer at one end and a ubiquitous sequence at the other (such as *Alu* repeat sequences). The steps that make up the cycles of the PCR method include: DNA denaturation, extension primer annealing, and amplification (or extension).

1. *DNA denaturation.* The double-stranded template DNA is denatured under high temperatures and the dissociated single strands remain free in solution.
2. *Extension primer annealing.* Two extension primers, which are selected by the sequence at the boundary of the region to be amplified, are utilized in order to anneal to one of the DNA strands. Each primer anneals to opposite strands; generally they are different in their sequence and are not complementary to each other. The primers are present in large excess over the DNA template, however, and this favors the formation of primer-template complexes at the annealing

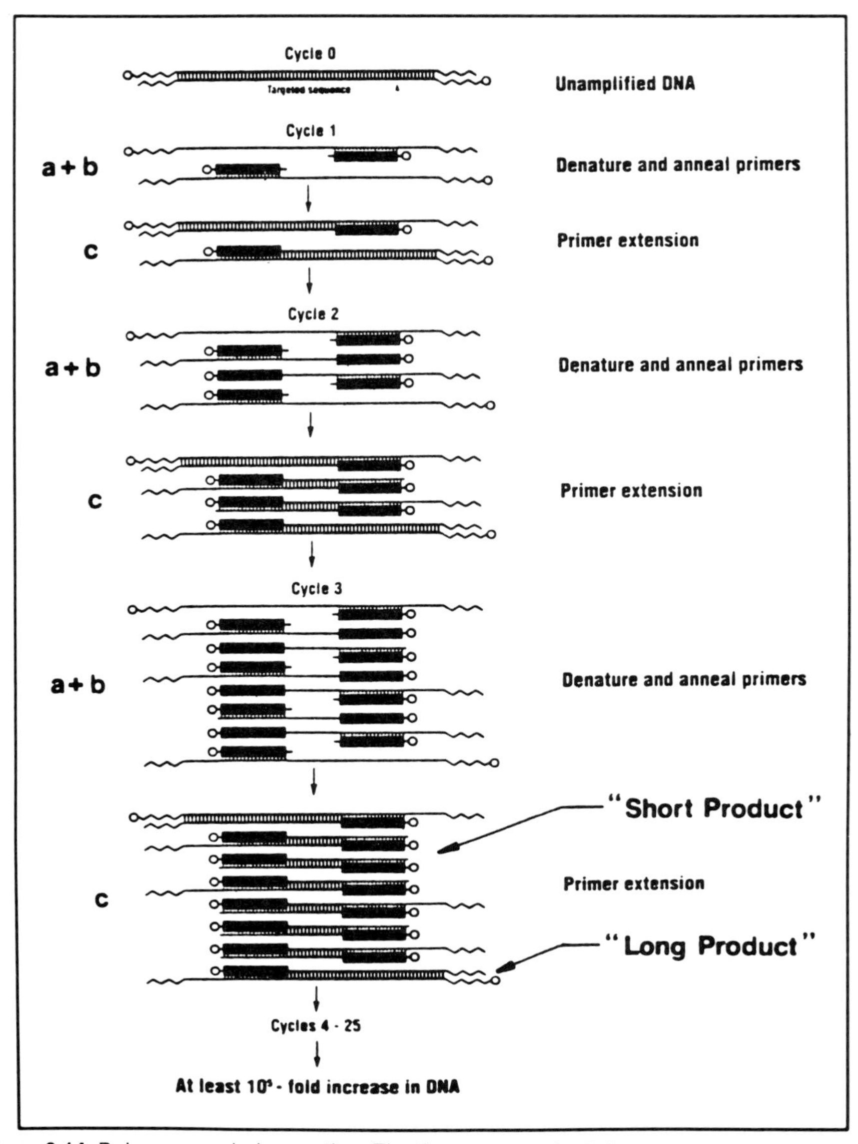

Figure 3.14 Polymerase chain reaction. The three steps to be followed once the primers have been added are: *denaturation*, which separates the double-stranded DNA into single strands and the primers from their respective complementary strands; *reannealing*, the phase during which the primers anneal to their respective complementary base pairs on each strand; and *extension*, the phase catalyzed by the enzyme polymerase during which the nucleotides are added to the primers, in one case in the sense direction and the other in the antisense direction. The cycle is repeated, and DNA has now been denatured so that the primers break away from the strands and the double strands separate into single strands. The primers again reanneal in their appropriate position, and one sees the stage is now set to develop proportionately more of the short segments of DNA. In cycle 3 one sees that the short DNA segments are being amplified preferentially over the long segments. This increases exponentially, so that by cycles 4 and 5, amplification of the short segments of interest predominates and the proportion contributed by the long segments becomes less and less important; thus, the main product at the end of 20 to 30 cycles will be the short DNA segment of interest. One increases the temperature to around 95°C for denaturation, but for reannealing the temperature is decreased to about 50°C, and during the extension cycle the temperature is increased to 72°C so the reaction proceeds rapidly. The DNA is again recycled (denatured, reannealed, and extended).

sites, rather than reassociation of DNA strands, when the temperature is lowered.

3. *Amplification (extension).* The 5′-to-3′ extension of primer-template complex is mediated by DNA polymerase and, as a result, the extension primers become incorporated into the amplification product. Initially the Klenow fragment of DNA polymerase was used for the reaction, but it was found to occasionally fail because of the high temperatures required. A thermostable DNA polymerase purified from the bacterium *Thermus aquaticus*, now known as *Taq* DNA polymerase (33), has gained wide usage and greatly simplified this process, since fresh enzyme now is not required after each denaturation step. This has allowed for automation to proceed.

With the wide acceptance of PCR, a variety of applications have been developed (34). A description of many of the newer applications of PCR is given next.

PCR Cloning

PCR has been successfully utilized for cloning (35), relieving the usual tedium found in the preparation of DNA fragments seen with classical subcloning methods. Modification of the 5′ ends of the extension primers allows unidirectional cloning into any vector without affecting the primers' ability to anneal specifically to the template. Additional sequences not complementary to the template and containing restriction recognition sites can be attached to the 5′ end of the extension primers during synthesis and subsequently be incorporated into the amplified product. This can later be separated from excess primers and the deoxynucleotide triphosphates and digested to generate the ends needed for subcloning.

Analysis of cDNA may also be enhanced by PCR. Once the first cDNA strand is synthesized, *Taq* polymerase can be added to promote second strand synthesis. Addition of a pair of specific extension primers allows amplification of specific cDNAs to proceed if the corresponding messenger was present in the initial mRNA. With this method, various tissues can be assayed for expression of the gene. Frohman et al. (36) recently devised a simple and more efficient DNA cloning strategy for obtaining full length cDNA clones of low-abundance mRNAs. Using PCR to amplify copies of the region between a single point in the transcript and either end (3′ or 5′), cDNAs are generated. The cDNA product may be generated in a single day and cloned, allowing production of large quantities of full-length cDNA clones of these rare transcripts. To use this protocol, known as rapid amplification of cDNA ends (RACE), a short stretch of sequence from an exon must be known, and from this region primers oriented in a 3′ and 5′ direction are chosen that will provide overlapping cDNAs when fully extended. The primers provide specificity to the amplification step. Extension of the cDNAs from the ends of messages to the specific primer sequences is accomplished by utilizing primers that anneal to the 3′ end or 5′ poly(A) tail. The overlapping 3′- and 5′-end RACE products are combined to produce an intact, full-length cDNA. This method provides an efficient alternative to other, more time-consuming cDNA cloning methods, as well as potentially being useful in the construction of cDNA libraries.

Another method of cDNA cloning by PCR using degenerate primers was developed and involves the synthesis of oligonucleotide probes based on known amino acid sequence. Prediction of the codon usage when designing these probes is difficult since the genetic code is degenerate. Lee et al. (37) described the novel synthesis of authentic cDNA probes based on known amino acid sequence using every possible primer combination coding for the amino acid sequence. The fundamental assumption behind this mixed oligonucleotide–primed amplification of cDNA (MOPAC) was that the authentic sequence primer would selectively anneal to its target complementary sequence, outcompeting less complementary primers during the annealing process (Fig. 3.15). Lee et al. showed that a perfect primer match is not necessary and that there is tolerance of up to 20% base pair mismatch between primer and template during the MOPAC reaction. Mixed primers of more than 1,024 combinations have successfully generated cDNA probes.

Inverse PCR (38) was developed because a major limitation of conventional PCR is that DNA sequences located outside the primer sequences are inaccessible. This is because an oligonucleotide that primes synthesis into a flanking region produces only a linear increase in copy number since no primer in the reverse direction exists. One purpose of inverse PCR is to allow in vitro amplification of DNA flanking a region of known sequence; inverse PCR utilizes the simple procedures of restriction digestion of the source DNA and circularization of cleavage products before amplification using primers synthesized in the opposite orientations to those typically used for PCR. In general, inverse PCR allows for amplification of upstream and/or downstream regions flanking a specified segment of DNA without resorting to conventional cloning procedures. This method can be used to rapidly produce probes for the identification, as well as the orientation, of adjacent or overlapping clones from a DNA library. This technique eliminates the need to construct and screen DNA libraries to walk thousands of base pairs into flanking regions and is particularly useful for determining the insertion sites of translocatable genetic elements and other repetitive DNA sequences.

The last PCR cloning method to be discussed is *Alu* PCR (39), which was developed to amplify human DNA of unknown sequence from complex mixtures of human and other species' DNAs. Previously, application of PCR to isolate and analyze a particular DNA region required knowledge of DNA sequences flanking the region of interest. Initially it was applied

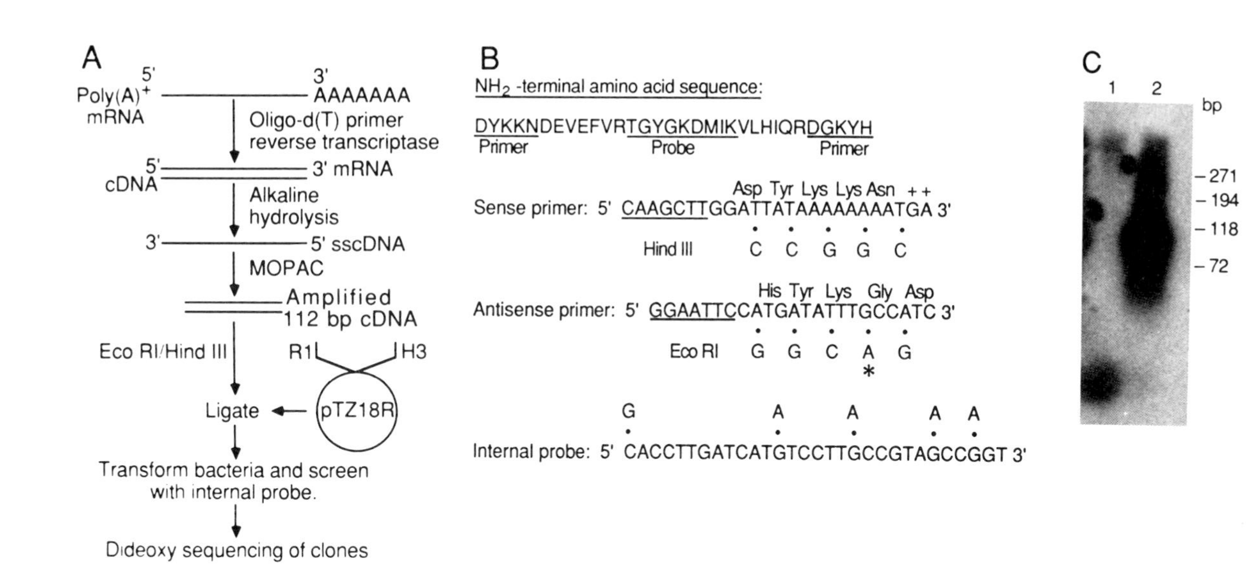

Figure 3.15 The strategy for MOPAC cloning, the selection of primers and probe, and product analysis. (A) Schematic steps in cloning cDNA based on amino acid sequence and the MOPAC procedure. (B) The NH_2-terminal amino acid sequence for porcine urate oxidase and the selection of primers and probe for MOPAC. The sense primer was synthesized to the amino acid sequence 1 to 5. The inclusion of the next two nucleotides from the sixth amino acid was indicated (+ +). The antisense primers were synthesized to the amino acid sequence 28 to 32. For both primers, every codon degeneracy was included except for the amino acid glycine, where the *asterisk* indicates the selected codon degeneracy. The selection of different restriction enzyme linkers (*Eco*RI/*Hin*dIII) is to facilitate the rescue of amplified cDNA. *Reprinted with permission from Lee CC et al. Science 1988;239:1288.*

to the isolation of human chromosome fragments in rodent cell backgrounds. This allowed isolation and characterization of sequences from specific regions of human DNA retained in the hybrids, obviating the need for cloned DNA libraries and isolation of human clones through the use of human-specific repeat sequence probes. *Alu* PCR has also proved useful for the rapid isolation of human insert DNA from cloned sources, including yeast artificial chromosomes. This technique utilizes the ubiquitous *Alu* repeat sequence found in human DNA. There are approximately 9×10^5 copies of this 300-bp sequence distributed throughout the human genome, with a known consensus sequence and regions of the repeat that are well conserved. PCR primers that are designed to recognize these conserved regions allow inter-*Alu* amplification for isolation of human DNA from complex sources.

PCR Sequencing

Another use of PCR is for nucleotide sequencing (40,41). The PCR method may be used as the initial step in sequence analysis, providing the generation of sufficient sample quantities for several subsequent analyses. If the region of interest is first amplified by PCR, the extension primers and deoxynucleotide triphosphates may be removed or replaced by a third primer (known as the sequencing primer) that is complementary to one of the strands of the amplified product, followed later by sequencing reactions. This *"triple primer" sequencing method* (Fig. 3.16) can be performed using the classical Sanger dideoxy sequencing conditions (27), incorporating one radiolabeled deoxynucleotide triphosphate, or alternatively using the third primer, radiolabeled at its 5′ end with a radioactive phosphate group. The "triple primer" method requires at least partial knowledge of the sequence being analyzed in order to synthesize the third primer. If no prior sequence exists, the same procedure may be used provided that the fragment to be sequenced is inserted in a vector of known sequence. Extension primers could be designed to anneal to the vector, resulting in an amplification product whose ends correspond to vector sequences. The "third primer" could be designed to anneal to the vector, thus initiating the sequencing reaction in the vector and moving into the insert.

Another sequencing method is *asymmetrical PCR* (42). Direct sequencing of the PCR product without an additional cloning step is generally preferred over sequencing cloned products. In addition to its simplicity, this procedure greatly reduces the potential for errors resulting from imperfect PCR fidelity, because any random misincorporations in an individual template molecule will not be detectable against the greater signals of the "consensus" sequence. Although several reports describing direct sequencing of double-stranded PCR products exist, protocols for

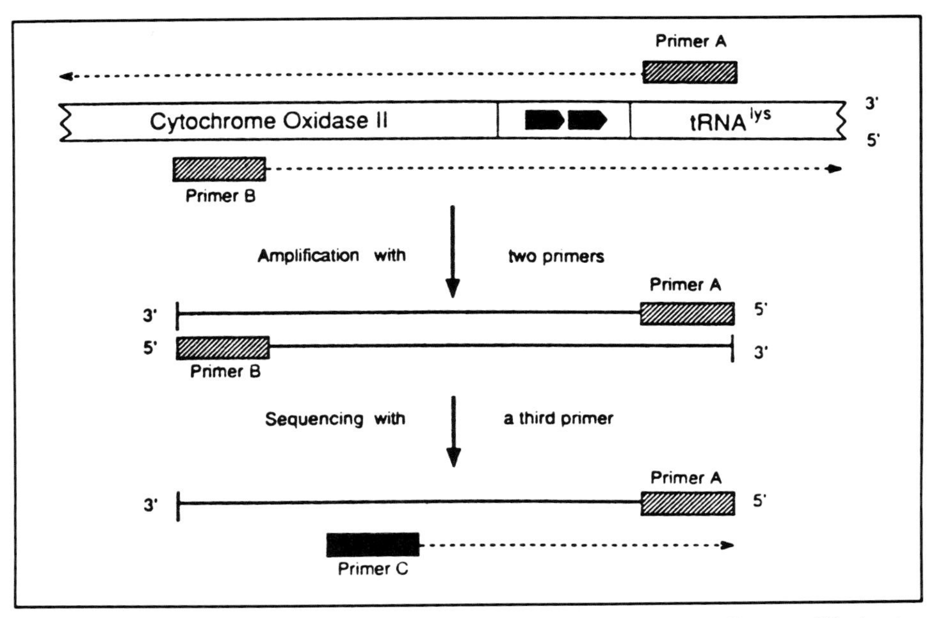

Figure 3.16 Triple primer sequencing method. The region of interest is first amplified using PCR. Primers A and B are the extension primers. After removing the extension primers and the deoxynucleotide triphosphates, the third (sequencing) primer C, labeled at its 5′ end with a radioactive phosphate group, is added and the sequencing reaction is conducted in the presence of mixture of deoxy- and dideoxynucleotide triphosphates. *Reprinted by permission from Wrischnik LA et al. Nucleic Acids Res 1987;15:529.*

preparation of double-stranded template DNA for sequencing were developed for covalently closed circular plasmids. Difficulties arise when this double-stranded sequencing protocol is applied to PCR-amplified fragments because of rapid reassociation of the short linear template strands. This may be avoided by modifying the PCR to produce single-stranded DNA of a chosen strand. This modified PCR uses an unequal (or asymmetrical) concentration of the two amplification primers. During the initial 25 cycles, most of the product generated is double stranded and accumulates exponentially. As the low-concentration primer becomes depleted, further cycles generate an excess of one of the two strands depending on which of the primers is limited. The single-stranded DNA accumulates linearly and is complementary to the limiting primer. The single-stranded template can be sequenced with either the limiting primer or a third internal primer, which provides an added degree of specificity. This method is less efficient than standard PCR and, therefore, more cycles are generally required to achieve a maximum yield of single-stranded DNA. Using 30 to 40 cycles typically gives the best results.

Techniques of Molecular Genetics

Reverse Genetics (Positional Cloning)

The biochemical bases of the vast majority of human genetic disorders are unfortunately not known, thereby increasing significantly the difficulty of gene mapping and cloning. The development of reverse genetics (43) makes it possible to localize the chromosomal position of the affected gene by the linkage methods described in Chapter 4. The overall objective with reverse genetics is to map the chromosomal location (locus) of the gene and to clone the region of DNA of interest, followed by isolation and identification of the gene. This is followed by ultimate sequencing and identification of the specific mutation responsible for the disease. Reverse genetics has been used in the isolation of several disease-causing genes, such as those for Duchenne muscular dystrophy (44), cystic fibrosis (45,46), retinoblastoma (47), and neurofibromatosis (48). The chromosomal localization of a gene involved in human disease has been accomplished by several techniques, which include cytogenetic detection of chromosomal abnormality, in situ hybridization of chromosomal spreads, and computerized linkage analysis (49–51). A more detailed map of the region of interest in the chromosome can be obtained by cutting and electrophoretically separating very large fragments of DNA from this region with a technique called pulsed-field gel electrophoresis. A method called jumping libraries, which is discussed later, has also been utilized to "jump" down the chromosome from an established marker toward the gene of interest. Once this region of the chromosome has been adequately defined, techniques capable of cloning large pieces of DNA have been used to develop one large fragment or several smaller overlapping fragments that span the region containing the gene. These cloned fragments may then be analyzed for the presence of exons encoding for the gene product expressed in the tissue of normal subjects and mutated in disease. One method is to screen cDNA libraries with the DNA fragments to find expressed RNA sequences in tissues that manifest the disease phenotype. Candidate clones of expressed sequence (cDNA) in these tissues may then be analyzed for the presence of expressed mutations that potentially correspond to the disease. DNA sequences can also be used to determine the amino acid sequence of the putative protein and develop antibodies. These antibodies can be used for immunohistochemical analysis of normal and diseased tissues.

Chromosomal Localization by Linkage Analysis

In those diseases in which neither the protein nor the genetic defect is known, one must first determine on which chromosome the locus for that gene resides (i.e., chromosome mapping). Since DNA is a monotonous, repetitive molecule of four bases, one must develop some land-

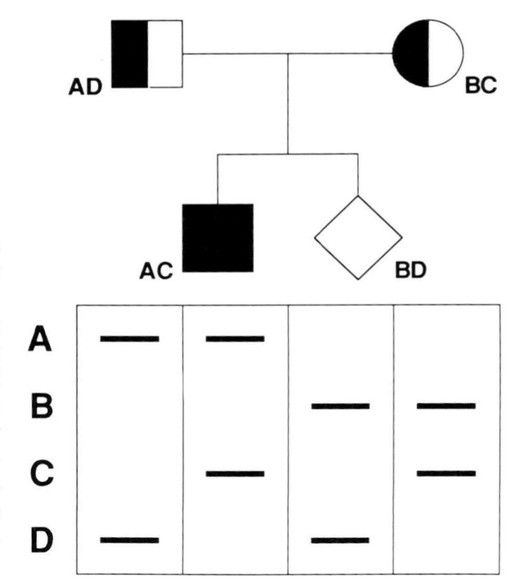

Figure 3.17 Types of DNA polymorphisms. (a) DNA polymorphism resulting from a single base substitution that eliminates a restriction site and yields a 7-kb rather than a 5-kb fragment (*left*) that is easily differentiated on a Southern blot (*right*). (b) Insertion/deletion DNA polymorphisms result from a different number of "tandem repeats" between two restriction sites, which in this case yield 6- or 8-kb fragments. (c) DNA polymorphisms resulting from presence or absence of a pseudogene, which changes the fragment lengths from 9 to 7 kb without affecting the recognition site.

marks. Fortunately, the appropriate location of certain DNA markers is known. The human genome has about 3 billion bp and there are approximately 600 known markers. If these were evenly spaced, there would theoretically be approximately 5 million bp between each of these markers, and therefore, linking the loci of a particular disease to one of them should be easy. However, this is not the case, since they are not evenly spaced in reality, with some DNA markers being separated by as many as 50 million bp. The DNA from family pedigrees is analyzed for various DNA markers utilizing these probes of known chromosomal location. The purpose of linkage analysis is to determine whether the locus of one of these DNA markers (on a known chromosome) and that of the disease in question are in close proximity, such that the marker and the locus for the diseased gene are coinherited more often than would occur by chance alone. The DNA markers are recognized by their polymorphism, detected on Southern blotting of the individual DNA, and whether the locus of the marker is coinherited with the disease locus is determined by computer analysis (Fig. 3.17) (see Chapter 4).

As previously noted, linkage analysis has been used over the past decade to localize the chromosomal location of many disease-causing genes. These include the X-linked disorders Duchenne muscular dystrophy (43) and fragile X-syndrome (52), as well as the X-chromosomal cardiac disorders of X-linked dilated cardiomyopathy (53) and cardioskeletal myopathy (Barth syndrome) (54). In addition, several autosomal cardiac disorders have also been localized, including hypertrophic cardiomyopathy (14q) (55,56), Romano-Ward long QT syndrome (11p) (57), and Marfan syndrome (15p) (58).

Transformed Lymphoblast Cell Lines

Lymphocytes are easy to obtain but, in addition, have the ability to be transformed into lymphoblasts, which means they can be grown in culture as an immortal cell line to provide a renewable source of that patient's DNA. The purpose of lymphoblast cell lines (LCLs) is then to develop a convenient source of renewable DNA. Whole blood (10 ml) is drawn from a patient, placed in tubes containing heparin, and immortalized by transformation with Epstein-Barr virus (EBV) and cyclosporin A (CSA). White blood cells are separated from the red cells by centrifugation through a ficoll gradient and washed twice with media (RPMI), and the cells are then adjusted to a concentration of 10^6 cells/ml of medium plus fetal calf serum, penicillin (1%), and streptomycin. CSA and EBV are then added, and the lymphocytes are cultured for 7 to 10 days and then fed with enriched media. These transformed cells may be used directly for DNA extraction or stored frozen to assure this renewable supply of high-quality genomic DNA (7). This allows for the long-term study of individual patient DNA without the need of constantly obtaining new blood from these individuals.

Restriction Fragment Length Polymorphisms

The discovery of DNA sequence polymorphisms by Kan and Dozy (59) has greatly facilitated the genetic analysis of humans. Examination of DNA from any two individuals will reveal DNA sequence variation involving approximately one nucleotide in every 200 to 500 bp of homologous genes (60). These polymorphic sequences occur much more frequently in DNA than in proteins (introns as well as exons), and most produce no deleterious clinical effect or destabilization of the DNAs. Some of these DNA sequence changes are detectable by restriction endonuclease digestion of the DNA (Fig. 3.18). When human DNA from normal individuals is digested with a particular restriction enzyme, fragments of discrete length are obtained. A single base pair change may abolish an existing recognition site for a restriction enzyme in the human genome or create a new recognition site, thereby altering the length of these fragments. Alternatively, since the number of tandem repeat sequences interspersed at various intervals in the human genome varies from individual to individual, when these repeat sequences occur between enzyme cleavage sites, the length of the DNA fragments generated by the enzyme digestion will vary. The term *restriction fragment length polymorphism* (RFLP) describes this fragment length variation that is generated by either mechanism (Fig. 3.19). The RFLPs have become useful as genetic markers to which loci for disease-related genes can be linked and then used to determine the chromosomal locus of the gene of interest. Since polymorphic restriction sites occur frequently on the human genome, a set of such sites can commonly be found in the region of the gene of interest.

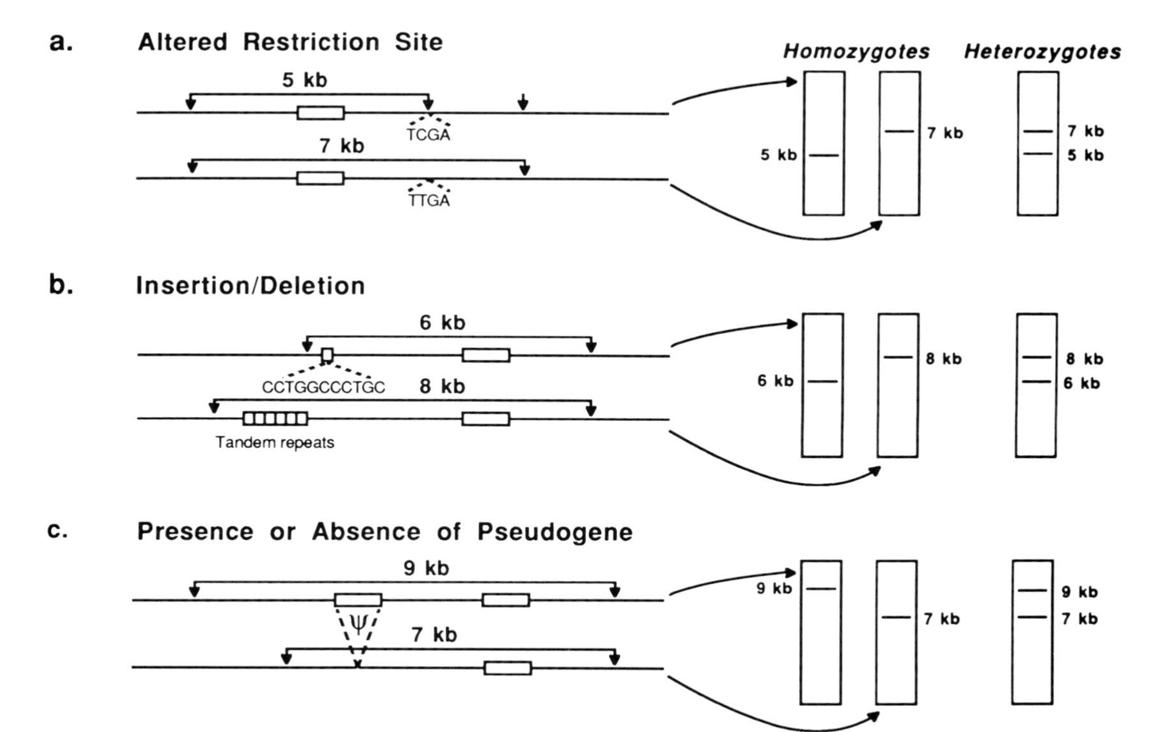

Figure 3.18 Linkage of an autosomal recessive disease gene to a DNA marker detected by a RFLP. Linked restriction fragments (bands) of different size cosegregate either with normal or with disease alleles in the family. On the basis of inheritance of the affected child (*solid square*), the father's disease allele cosegregates with band A and the mother's diseased allele with band C. The normal offspring (indicated by an *open diamond*) inherited the father's B and the mother's D bands, both of which are linked to the normal gene; hence, the offspring is homozygous normal.

To detect these RFLPs one performs a Southern blot. The patient's DNA is obtained immediately, as indicated previously, from leucocytes or from lymphocytes grown in culture and digested with appropriate endonuclease restriction enzymes. The DNA fragments are separated by electrophoresis, denatured, and transferred to a filter for hybridization. After the radiolabeled probe is hybridized overnight with the filter, it is washed and placed in an x-ray cassette prior to being put into a −70°C freezer to allow the radioactive bands to appear on the x-ray film. The bands of the autoradiogram may be the same in all individuals (constant bands) or may be different, (polymorphic). Performing this analysis to determine whether a set of such sites is present around the gene for chromosomal localization will also provide the opportunity to classify chromosomes into their different haplotypes as well. These chromosome haplotypes (i.e., combination of alleles from closely linked loci, usually

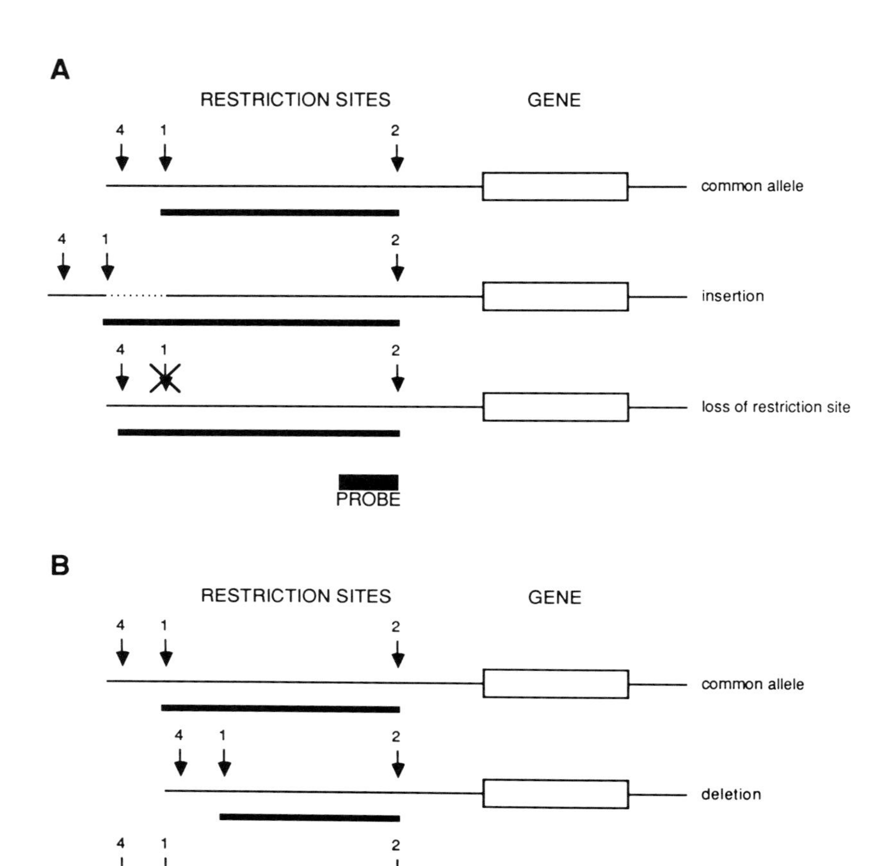

Figure 3.19 (A) Molecular basis of RFLP (smaller fragments). Compared with the common allele, smaller fragments may be generated by deletion of DNA between two restriction sites (here, 1 and 2) or by creation of a new site (here, site 3). *Reprinted by permission from Ostrer H, Hejtmancik JF. J Pediatr 1988;112:679.* (B) Molecular basis of RFLP (larger fragments). Using a particular probe, a common restriction fragment (or allele) will be found in the population (here, generated by sites 1 and 2). Compared with this allele, larger fragments may be seen. These are shown as *bars* in the figure. Larger fragments arise from insertion of DNA between two restriction sites (1 and 2) or from loss of site (here site 1).

with some functional affinity, found on a single chromosome, and segregating together with a particular trait) are useful in marking the chromosome at a gene locus and also for tracing the origin and migration of genes.

Most of the polymorphic DNA markers for human chromosomes presently available have only two alleles. While two-allele RFLPs can be useful for linkage, they usually are not as informative as systems with multiple alleles. Fortunately, a small number of known markers, known as *variable number of tandem repeat markers* (or VNTRs), detect loci that produce fragments having many different lengths when digested with restriction enzymes. This polymorphism occurs secondary to variations in the number of tandem repeat sequences in that short DNA segment. Since most individuals are heterozygous at these loci, VNTRs can potentially provide linkage information in a large number of families. Nakamura et al. (61) produced a series of single-copy probes from oligomeric sequences derived from tandem repeat regions of a variety of genes and showed them to be highly polymorphic. These sequences have become very valuable for linkage analysis and mapping of genetic disease loci over the past several years.

Isolation of a Gene

The techniques described above are classic routine methods for molecular biological research. Recently, a series of techniques have been developed in order to move across chromosomes more rapidly. These are known as megabase methods for chromosome cloning. Until the mid-1980s, the size of DNA fragments that could be analyzed or cloned severely restricted the distances that could be covered by chromosome walking, and the fragments were much too small to allow for in-depth studies of these complex genomes. The total length, in base pairs, of the human haploid genome is approximately 3×10^9, and the smallest distance measurable by recombination analysis (linkage analysis) is at best on the order of 1 centimorgan (cM), or 1% recombination, which corresponds to about 1,000 kilobase pairs (kb). While this is only a mean value, it indicates the scale at which genetic linkage analysis operates. Cytogenetic analysis, whether based on translocations, deletions, or in situ hybridization, typically can resolve up to a chromosome band. Since the total number of individualized bands in human chromosomes is approximately 800, the best resolution obtained, utilizing this method, is on the order of 4,000 kb. As described earlier, classical recombinant DNA technology can only deal with much smaller pieces of DNA. Fractionation by agarose gel electrophoresis resolves DNA molecules up to 30 to 100 kb long, with larger molecules unable to be separated effectively. The effective size range studied by Southern blotting is 20 to 30 kb around the sequence homologous to the probe utilized. Similarly, cosmid cloning, which is amenable to the largest inserts of DNA compared to plasmids (less than 15 kb) or bacteriophage (less than 25 kb), permits DNA inserts of 40 to 45 kb; nevertheless, chromosome walking strategies based on reiterative screening of genomic libraries proceed with "steps" of approximately 20 kb. These are both time consuming and labor intensive.

This large difference between the distance covered by cytogenetics and linkage analysis genetics on the one hand and cloning and analytical techniques on the other leads the researcher to a significant level of frustration. Since many of the problems of human genetics are those that require moving from point A (the cloned gene or DNA sequence shown by linkage analysis to lie quite close to point B) to point B (the location of the gene whose dysfunction is responsible for a particular inherited disease), better methods became necessary. This need stemmed from the fact that "quite close" usually meant 3 to 5 cM or 3,000 to 5,000 kb, or in the same chromosome band well outside the range of conventional blotting, cloning, or even walking techniques. The following methods have recently helped to overcome many of these problems.

Chromosomal In Situ Hybridization

Various strategies are available for the localization of cloned genes and random DNA sequences within the human gene. One of the best described of these methods is in situ hybridization, first described by in 1969 by Gall and Pardue (62). This technique involves hybridization to a panel of DNA obtained from somatic cell hybrids with different human chromosomes present. For regional localization within a particular chromosome, sequences can be hybridized to the DNA from cell lines carrying deletions of that chromosome, and thereby mapped within or outside the deletion. A panel of somatic cell hybrids with a variety of rearrangements of a particular chromosome can also be used in this way. The smallest region of overlap that gives positive hybridization signals can then be determined. In situ hybridization provides a direct approach to regional mapping by hybridization of known nucleic acid sequences to their complementary DNA within fixed chromosome preparations.

In situ hybridization has been used during the past two decades to localize sequences repeated many times within the human genome. Until the recent development of DNA recombination technology, the paucity of pure probes stood as an obstacle for localization of other specific gene sequences. Now that higher quality probes are available, and with recent improvements in hybridization, efficiency, and chromosomes banding, major improvements in signal resolutions have emerged, permitting localization of single-copy sequence DNA and repetitive sequences. This method has therefore become an important complement to the mapping of the human genome.

Chromosome Walking

Localization of the chromosomal locus of a gene by linkage analysis indicates that the gene locus is linked to the marker locus, which means anywhere from about 1,000 kb to 20,000 or 30,000 kb. Even if the two loci are within 1,000 kb, this is too large for conventional cloning, so one attempts to identify markers closer on either side of the gene locus that are less than 1,000 kb away. The techniques used in restriction mapping and subcloning of overlapping clones of the genomic DNA are referred to as chromosomal walking. This technique (63) can be applied to isolate gene sequences whose location is known but whose function is not. First, cloned genomic sequences are localized in the genome by radiolabeling the clone fragment and then using it as a probe to isolate other clones. The radiolabeled probe may also be used, if necessary, in a hybridization experiment in situ with cytological preparations of chromosomes. In this way, a random set of cloned genomic DNA is localized and one is chosen whose location on the chromosome in question is closest to the mapping position of the locus of the gene under investigation. Screening of a genomic library with this clone as a probe ensures that other overlapping clones that contain the appropriate DNA sequences are selected. The overlap can be on either side (5′ or 3′) of the probe. Repetition of this step will further select clones with sequences complementary to these overlaps, thus extending the cloned regions (Fig. 3.20). Such walking will occur in both directions along the chromosome unless there is some means of distinguishing the direction. Certain libraries, such as λdash, allow directional cloning. Repetitive DNA sequences are known to be dispersed throughout the genome, and these can be troublesome when screening libraries. For this reason, the probe must be unique in sequence. Once the locus in question is near, sequence comparisons should be possible by direct means or by restriction mapping. Using the walking method, entire genomic sequences, together with substantial flanking sequences, may be obtained with single steps in either direction.

Chromosome Jumping Libraries

Chromosome walking makes use of overlapping phage or cosmid clones to progressively obtain sequences distal to a given starting point. The problem with this approach lies in the fact that each step that may extend the existing map of the region of interest by 40 kb (the cosmid insert size) at best involves screening the complete genomic library, restriction mapping of the resulting clones, and obtaining a few new single-copy probes from the end of choice. This makes the approach very labor intensive. Chromosome walks along several thousand kilobase pairs are, in principle, required to cover the distance between human loci separated by a few centimorgans, and walks such as these are not feasible. Every sequence between the start point of any walk and the destination point, possibly several thousand kilobase pairs away, must be cloned and characterized in this method. This may be unnecessary if only the destination point is of interest, however. Jumping libraries were developed to avoid most of these unnecessary steps and only touch the chromosomes at widely spaced intervals or jumps. The basic feature of these methods is the circularization of large DNA fragments, bringing together the two ends

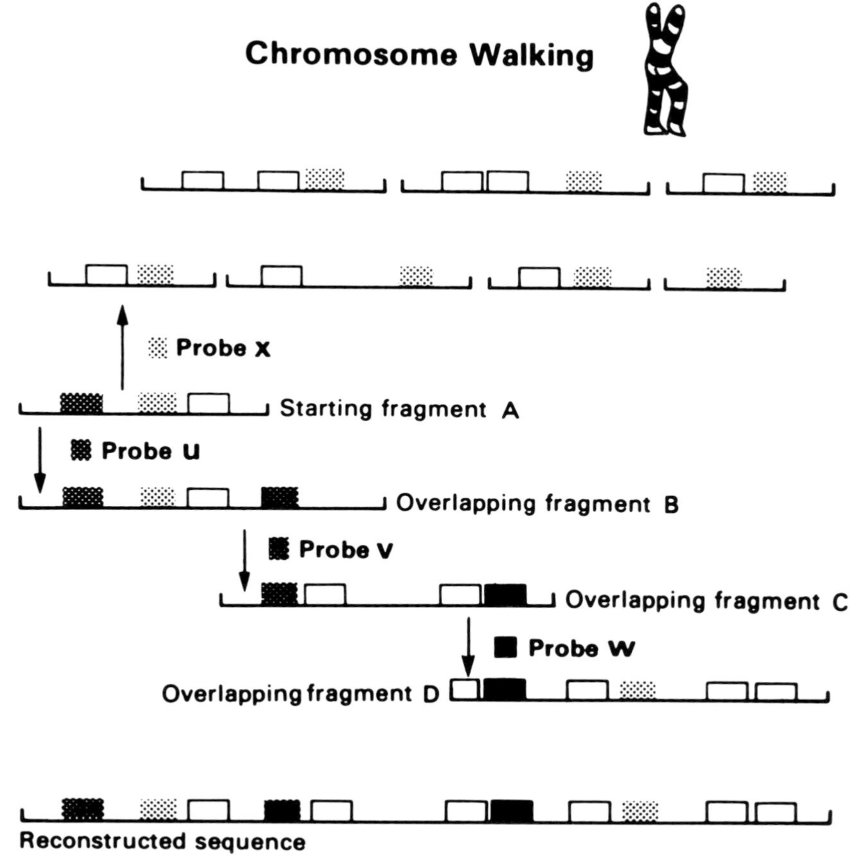

Figure 3.20 Chromosome walking from point A to point B on the chromosome using restriction fragments that have been cut from the region of interest and cloned. To arrange the fragments in their proper order one must find overlapping DNA fragments that have matching sequences, such as *fragments A* and *B*. To find the matching sequences, a small radioactive probe (*probe u*) is prepared from starting DNA fragment *A*. If the probe is unique, it will hybridize to the unique overlapping fragment *B*. The unique probe from fragment *B* (*probe v*) can then be used to find contiguous *fragment C*. If the probe is not unique, such as *probe x*, it will not identify one specific fragment and will not be useful for sequence reconstruction. By repeating these steps with new unique probes from each overlapping sequence, the order of the entire region of the chromosome can be determined. Picture several copies of one manuscript page randomly cut into pieces of paper, each containing five to six words. Finding the same unique word on two different paper fragments would permit reconstruction of the order of two fragments. *Reprinted with permission from Schmickel R. J Pediatr 1986;109:231.*

of the fragment. If performed in the presence of a selective marker, subsequent steps may result in a library in which each clone contains essentially the two ends of the large DNA fragment from which it was derived. In general, jumping libraries should avoid the pitfalls of repetitive regions that plague chromosome walking, since these regions can now be jumped over.

Two approaches have been developed, one in the laboratory of Lehrach (64), and the other by Collins and co-workers (65). The Lehrach approach was one in which the large DNA fragments are generated by complete digestion of the cellular DNA with a "rare cutter" enzyme such as *Not*I (Fig. 3.21). The jumping library created contains in each clone the two ends of a given *Not*I fragment. In order to utilize this library, a starting probe is required that is located adjacent to a *Not*I site. To obtain this probe, however, a chromosome walk from the start point to the first *Not*I site may be required. Screening the library with this probe will then pull out a clone containing the other end of the *Not*I fragment, a few hundred kilobase pairs from the starting point. It is next necessary to cross the *Not*I site and obtain a probe located on the other side of this site. This can be done by screening a standard genomic library or a specially constructed smaller "jumping library" as a "junction library" with the probe obtained previously. The procedure can then be repeated, with each jump providing a new probe located several hundred kilobase pairs further from the start point, depending on the length of the particular *Not*I fragment. Two drawbacks to this approach are notable: first, the need for a start probe close to the enzyme site, which adds the necessity for chromosome walking; and second, and more serious, the problem of missing clones.

A more general implementation of the jumping library principle is the method developed by Collins et al. (65). In this case, the large DNA fragments are obtained by partial digestion with a "frequent cutter" enzyme such as *Mbo*I, and the fragments are subsequently sized by preparative fractionation of pulse-field gels before circularization and cloning of the ends, as previously noted. The resulting library must be quite large (several million clones) to be representative, but can be used with any starting probes since all sequences should be represented. Since the method is based on circularization of a collection of fragments of similar length, it should not be interrupted by the kinds of gaps found in the Lehrach method. Jumps of greater than 100 kb may be performed using this method. The principle behind this method depends on formation of large genomic circles from size-selected DNA, bringing together genomic fragments that initially were far apart. Partial *Mbo*I digestion is performed prior to size selection, causing no significant bias for a particular sequence to occur at the circle junctions. In each clone, the position of the joining fragments is marked by the suppressor tRNA gene, which allows for selection of these fragments after restriction enzyme digestion of the circles. These are then ligated into a phage vector. This technique has become a useful and powerful cloning method, and recently the gene causing cystic fibrosis was cloned with the aid of this technique (65).

Pulsed-Field Gel Electrophoresis

The pulsed-field gel electrophoresis (PGFE) technique was originally developed by Schwartz and Cantor (66) and later modified by Carle and Olson (67). It allows for improved resolution of large DNA fragments up

Chromosome Jumping and Linking with Not I

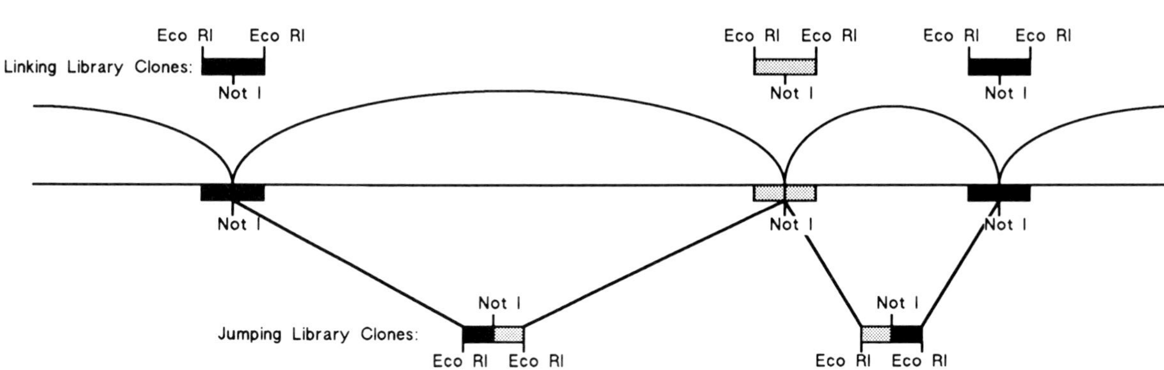

Figure 3.21 Chromosome jumping is a procedure used to develop more closely flanking markers to the region that is believed to contain the gene of interest and, ultimately, to clone the complete intervening DNA. Using *Not*I restriction enzymes, whose recognition sites occur very infrequently, one cuts the DNA of interest into very large segments. These segments are then ligated together into circular DNA together with an appropriate restriction site, *Eco*RI, and a reporter gene so that they can be recognized and subcloned. One can move rapidly across the chromosome at a rate of 100 to 200 kb instead of the conventional 20 to 40 kb; the latter process is referred to as chromosomal walking.

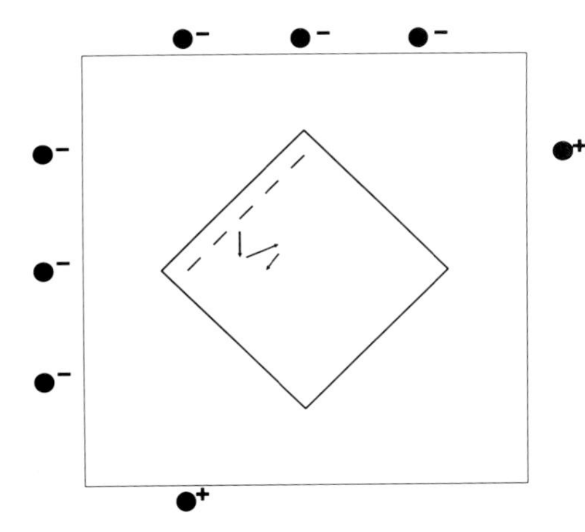

Figure 3.22 Pulsed-field gel electrophoresis (PFGE). The current is applied alternatively in different directions, as illustrated here.

to 2×10^6 bp, or 2 megabase paris (mb), because of the near linearity of the separation, with particularly good resolution at the larger fragment size. In principle, length differences of 10 to 20 kb are detected among fragments ranging from 100 to 800 kb in size. DNA from entire chromosomes of lower eukaryotes and megabase pair–sized DNA from higher eukaryotes have been separated by PFGE. DNA molecules of 20 kb or smaller, upon electrophoresis, move at a rate proportional to their size, whereas DNA molecules greater than 20 kb exceed the size allowed by the gel matrix. These molecules can still move by deforming their shape, but the velocity of migration is essentially the same for all large molecules since there is no sieve effect and, thus, there is no separation. In conventional electrophoresis, the electrical field is applied constantly in one direction. In PFGE, the electrical field is applied alternatively in two directions, which forces the molecules to reorient and move in two different directions (Fig. 3.22). Larger molecules spend considerable time reorienting, whereas smaller molecules reorient faster and begin to move such that effective size fractionation can occur. The time the electrical field spends in each direction is referred to as the pulse time. The optimal pulse time varies with the size range of molecules to be separated, such that larger molecules require larger pulse time.

Because of the separation range of PFGE, restriction enzymes must be selected that cleave infrequently. The enzymes with the widest application in normal gel electrophoresis recognize hexanucleotide or smaller sequences and produce fragments averaging 3 kb ranging from 10 bp to 50 kb. Only a handful of commercial enzymes, such as *Not*I and *Sfi*I, recognize octomeric or longer sequences. An endonuclease recognizing 6-bp or 4-bp sequences would produce fragments averaging 4,096 and 256 bp, respectively. Enzymes containing multiple CpG dinucleotides in their recognition site also appear useful since these sequences are underrepresented in the genome by one order of magnitude. PFGE has become useful for the location of genes responsible for the genetic defects in humans by assisting in the preparation of physical maps of regions separated by megabase pairs. Molecular genetic analysis of genomes has therefore been simplified using PFGE (Fig. 3.23).

Yeast Artificial Chromosome Cloning

Prior to the development of yeast artificial chromosome (YAC) cloning, the best cloning system available utilized cosmid vectors. Since the entry of cosmid DNA into bacterial cells involves λ phage particles, the absolute upper limit of size of the cosmid is the length of DNA that can be packaged in a phage particle (i.e., 50 kb) of which approximately 5 to 10 kb is used up by the vector sequences. Therefore, the largest DNA fragments that can be cloned by cosmids are only 40 to 45 kb long. YAC cloning was developed in an attempt to overcome this size limitation.

Burke et al. (68), in 1987, first successfully implemented the YAC system for cloning very large human DNA segments. Development of YACs involves isolating centromere and telomere sequences, preparing a yeast DNA replication of the original DNA, and then inserting a larger piece of human DNA with selective markers. This makes a YAC that could be stably introduced into yeast spheroplasts and grown as a yeast colony, propagating the human DNA as one of its own chromosomes (Fig. 3.24). Such artificial yeast chromosomes have been shown to allow stable propagation of 100 to 700 kb of human DNA fragments. This allows for the possibility of construction of complete libraries of human DNA in yeast and the cloning of a gene or region of interest as one, or a set, of such large fragments. These fragments can subsequently be mapped in detail or used to prepare many libraries. If the mean size of inserts contained in the library is 500 kb, a complete library need only contain 10,000 to 20,000 clones to cover the genome several times, with each clone being able to provide a wealth of information. Despite the beauty of this system, some technical problems have limited its usefulness. Handling of large DNA fragments in solution is difficult and may limit the maximum insert size. Many investigators have attempted to handle these large DNA fragments in agarose gels to overcome this problem. In addition, the efficiency of yeast protoplast transformation is low, giving hundreds of recombinants per microgram of transforming DNA versus the 10^5 to 10^7 recombinants produced with phage cosmid or plasmic vectors in *E. coli*. This makes the creation of complete human libraries a task at which few laboratories have succeeded.

The screening of yeast libraries is less straightforward than that of phage, plasmid, or cosmid libraries. Technically the procedure of yeast colony hybridization on filters is similar to the *E. coli* procedure, except that prior overnight incubation of filters with zymolase has been required to digest the tough cell wall so that DNA can be released from the cell.

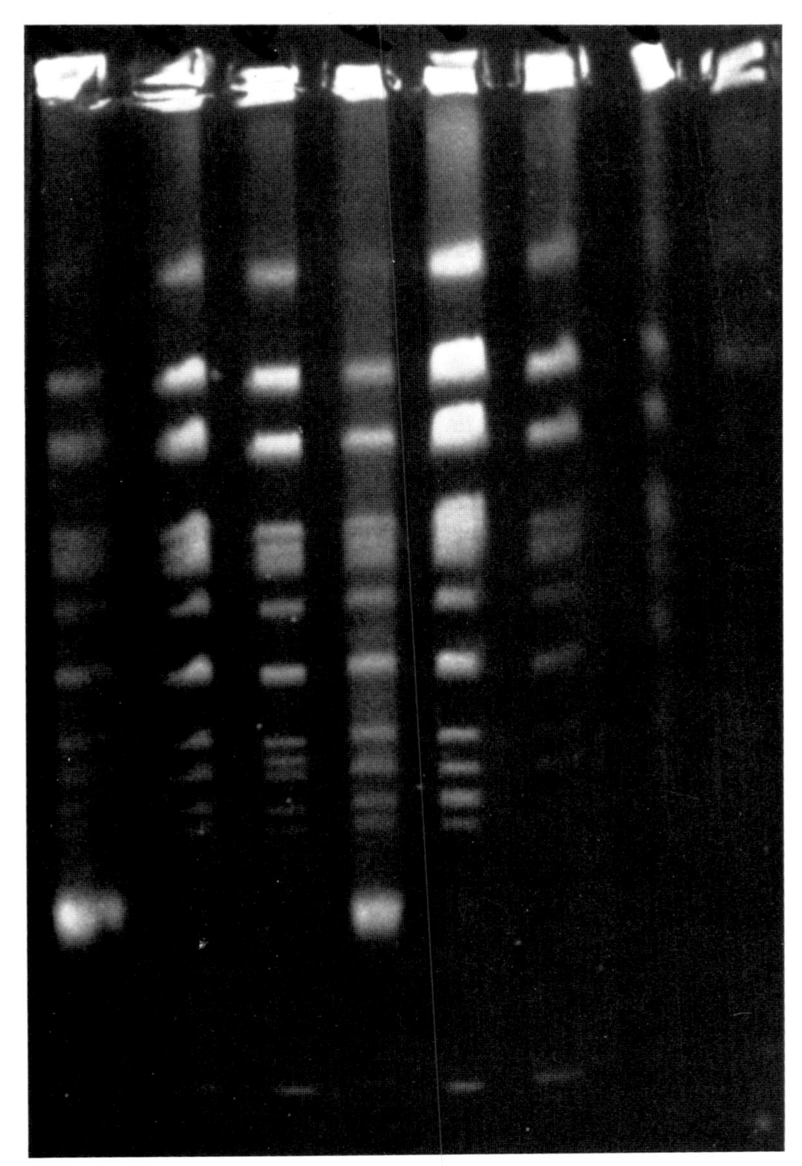

Figure 3.23 Pulsed-field gel. Shown here are large segments of DNA following separation by PFGE. Each vertical lane represents a separate DNA sample that was obtained from a yeast artificial chromosome library with a human insert. Each fragment of DNA exceeds 200,000 bp. The separation of large fragments of DNA is as distinct with this technique as that of small fragments (less than 20 kb) would be after conventional electrophoresis on agarose. The bands are visualized utilizing ethidium bromide staining.

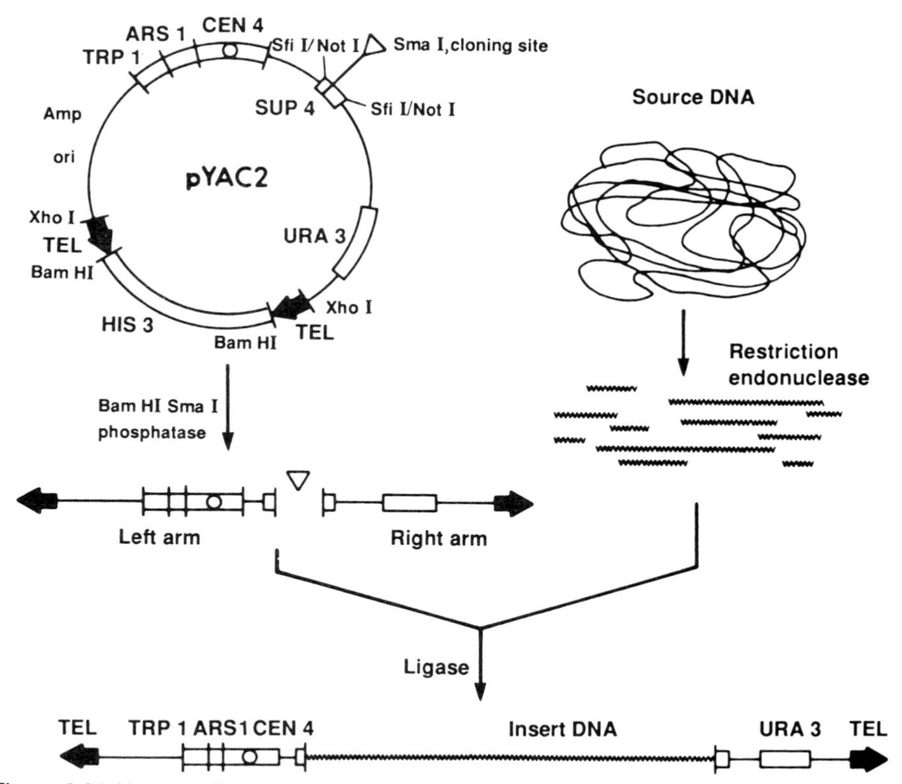

Figure 3.24 Yeast artificial chromosome (YAC) cloning system. Construction of a YAC consists of inserting the human DNA of interest into a YAC vector. However, there must be several key elements in the YAC vector to make this possible and to develop an artificial chromosome. The key regions of the pYAC vector are as follows: TEL (*black arrows*) are the yeast telomeres; ARS 1 refers to the autonomous replicating sequence, which is essential for replication; CEN 4 is the centromere from yeast chromosome 4; TRP 1, URA 3, and HIS 3 are yeast marker genes; AMP is the ampicillin resistance from pBR 322; and ori is the origin of replication of the pBR 322. *Bam*HI, *Sfi*I/*Not*I, and *Xho*I are recognition sites for these respective restriction endonucleases. The *Sma*I recognition site for that endonuclease is the cloning site, and the vector is cut with *Bam*HI and *Sma*I phosphatase. This cleaves the pYAC2 into a linear molecule with a left arm and a right arm. The DNA, which is cut with an appropriate restriction endonuclease, is ligated into the cloning region in the presence of ligase to give the completed YAC. The human DNA of interest is labeled "Insert DNA." The overall vector now has two telomeric ends like naturally occurring chromosomes. The main advantage over conventional cloning is that the human DNA insert is much larger and can be up to 1,000 kb.

Another problem is detection of the yeast clone. Cloning in bacteria, amplification of plasmid or cosmid, or lytic replication of phage all result in colonies or plaques in which the cloned DNA represents a large proportion of the total DNA present. Detection of the signal produced by hybridization of a probe is then no problem. Yeast clones, contrast, contain just a single copy of the artificial chromosome in the total yeast genome (about 15,000 kb), and therefore the signal obtained from a typical probe 1 kb long may be difficult to detect among background hybridization on all this DNA. Separation of the YAC by PFGE prior to Southern blotting allows more reliable detection of a unique sequence from the YAC clone. During the past few years several laboratories have been successful in cloning human genes by utilization of YAC libraries. As the number of laboratories using the St. Louis library (developed by Olsen) or those developing other libraries grows, the likelihood of this technology leading to the cloning of larger numbers of disease-causing genes increases dramatically.

Restriction Mapping

A map of restriction enzyme recognition sites of DNA may be prepared by restriction endonuclease digestion of that DNA using a series of different enzymes. This map may be used, for example, in defining the introns and exons of a transcript; if both genomic and cDNA clones are available, a comparison of the two may give an approximate indication of the location of exons and introns. This may be confirmed by Southern blot analysis of the genomic DNA with the cDNA (69).

Detection of Mutations

There are three basic methods that may be used to detect mutations: 1) the chemical cleavage method, 2) denaturing gradiant gel electrophoresis, and 3) RNAse protection assay. PCR may be used to amplify the area to be analyzed by any of these techniques.

Chemical Cleavage

The chemical cleavage method, described by Cotton and colleagues (70), takes advantage of the development of heteroduplexes when thymine (T) is mismatched with cytosine (C), guanine (G), and thymine or when cytosine is mismatched with thymine, adenine (A), and cytosine. They showed that these heteroduplex DNA sequences, when incubated with either osmium tetroxide for the T and C mismatches or hydroxylamine for the C mismatches, followed by piperidine incubation, cleaved the DNA at the modified mismatched base. Using end-labeled DNA probes that contain T or C single base pair mismatches, Cotton and colleagues showed that cleavage was at the base predicted by sequence analysis. Their procedure detects all types of mutations, including insertions, deletions, and base changes.

Denaturing Gradient Gel Electrophoresis

Another method for detecting mutations uses GC clamps and denaturing gradient gel electrophoresis (DGGE). Denaturing gradient gel electro-

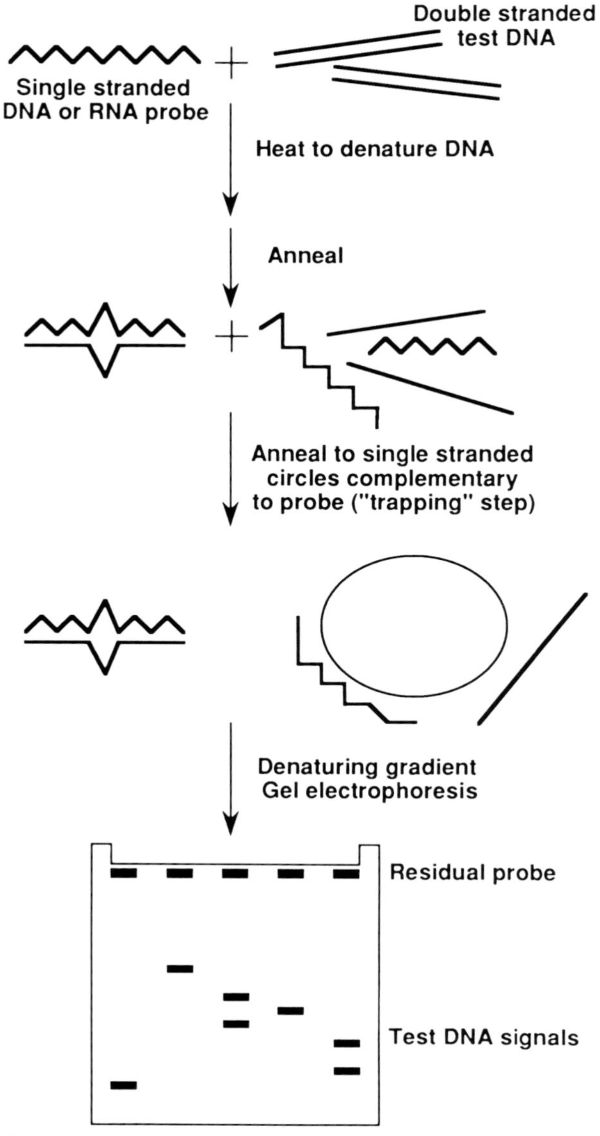

Figure 3.25 A scheme for detecting single base substitutions by DGGE. A labeled single-stranded DNA or RNA probe is mixed with double-stranded cloned, genomic or PCR-amplified DNA, and the mixture is heated to separate the DNA strands. An annealing reaction is performed, resulting in hybrid duplexes between the probe and its complementary strand in the test DNA sample. If a mutation is present in the test DNA sample, a single base mismatch will be present in the probe:test strand duplex. After the annealing reaction, an excess of single-stranded circular DNA complementary to the probe is added to the mixture to bind any residual probe. The mixture is then subjected to electrophoresis in a denaturing gradient gel. The gel is dried and exposed to x-ray film. Mismatches in mutant DNA samples that alter the melting behaviour of the fragment cause the fragments to shift upward in the gel due to their decreased thermal stability. Note that this procedure takes advantage of the increased resolution on DGGE of mismatched molecules. *Reprinted with permission from Rickwood D, Hames BD (eds): Genomic Analysis: A Practical Approach. Washington, DC: IRL Press, 1988, Myers RM, Sheffield VC, Cox DR, p. 104.*

phoresis (Fig. 3.25) allows separation of DNA molecules differing by as little as a single base change. This separation is based on the melting properties of DNA molecules in solution. DNA molecules melt in discrete segments, called melting domains, when the temperature or denaturant concentration is raised. Melting domains vary in size from 25 bp to hundreds of base pairs, and each melts cooperatively at a distinct temperature (T_m). The T_m of a melting domain is highly dependent on its nucleotide sequence because of stacking interactions of the bases. Small sequence changes can lead to large T_m changes. In DGGE, DNA fragments are electrophoresed through a polyacrylamide gel containing a linear gradient (top to bottom) of increasing DNA denaturant concentration. As the DNA fragment enters the concentration of denaturant where its lowest T_m exists, the molecule forms a branched structure with retarded mobility in the gel matrix; hence, the fragments separate. DGGE can be used to detect single base changes in all but the highest T_m of a DNA fragment, the latter being impossible to detect because of the loss of sequence-dependent migration of fragments upon complete strand dissociation. This can be overcome with cloned DNA fragments by attaching a GC-rich segment, known as a "GC clamp," to a DNA fragment that melts in two domains. In the absence of the GC clamp, only those single base changes that lie in the first melting domain of this DNA fragment will separate by DGGE. Attaching the GC clamp allows the separation of essentially all single base changes that lie in the second domain. A 30- to 45-bp GC clamp can be attached to DNA fragments for PCR, allowing detection of a single base change in the attached DNA fragment during DGGE. The huge amplification of DNA fragments by PCR during this procedure increases the sensitivity, so that very small amounts of genomic DNA are required; signals are detectable by ethidium bromide staining.

RNAse Protection Assay

RNAse A is an endoribonuclease that specifically recognizes single-stranded RNA 3′ to pyrimidine residues and cuts the phosphate linkage of these bases to the adjacent nucleotide. This produces pyrimidine 3′ phosphates and oligonucleotides with terminal pyrimidine 3′ phosphate (71). Myers et al. (72) and Winter et al. (73) used the effects of this enzyme to develop a method for mapping single-base mutations in DNA or RNA, where single-base mismatches in RNA-DNA or DNA-RNA hybrids are recognized and cleaved by RNAse A. Next a ^{32}P-labeled RNA probe that is complementary to wild-type DNA or RNA is synthesized and annealed to test DNA or RNA containing a single-base mismatch, and the location of this mismatch is determined by analysis of the sizes of cleavage products using gel electrophoresis. This method detects up to 50% of all possible single-base mismatches.

S1 Nuclease Mapping

S1 nuclease mapping usually involves hybridization (74) of nuclear RNA or mRNA with a genomic clone fragment under conditions favoring formation of DNA-RNA hybrids (38) rather than DNA-DNA hybrids. Any single-stranded DNA not hybridized with the RNA is then digested with S1 nuclease that is specific for single-stranded DNA. Hybridization with its complementary RNA protects the surviving DNA from digestion, and this DNA can be analyzed on a gel to determine its size (Fig. 3.26). Alternatively, the protected strand of DNA can be subjected to DNA sequence analysis to determine the exact ends of binding to the RNA. This method can determine the site of the 5′ end of RNA, the presence of introns at the 5′ end of genes, or the sites of insertions or deletions in the RNA molecule.

Haplotype Analysis by PCR

PCR has been adapted for use in RFLP and haplotype analysis. A haplotype is a combination of alleles at two or more loci on the same chromosome. As a result of selective forces or short physical distances between loci, combinations of alleles at different loci on the same chromosome may be inherited as a unit. Previously, analysis of the combination of alleles on individual chromosomes was feasible only by reconstruction of parental chromosomes from the segregation of alleles in pedigrees. Development of PCR haplotype analysis has made possible the genetic analysis of individual chromosomes without resorting to pedigree analysis. Combinations of alleles along a chromosome can be determined from a small number of sperm or single diploid cells using DNA-based typing of regions amplified in vitro by PCR. This method can also be used to study the recombination frequency between loci that are too close for pedigree analysis to yield statistically reliable recombination rates. RFLP analysis using PCR has been accomplished by use of PCR primers from specific regions or repeat sequences to amplify DNA from family members. Recently, dinucleotide repeat polymorphisms or sequence-tagged sites have been developed, giving rise to PCR-based VNTRs (Fig. 3.27). Most of these approaches do not require radioactive labeling, being able to rely on ethidium-stained gels only. One of the more common uses of PCR RFLP analysis has been that of human lymphocyte antigen typing. More recently, RFLP analysis using PCR has been adapted for the duchenne muscular dystrophy region of the X chromosome. RFLP analysis using PCR allows for more rapid linkage analysis than is allowed by classical Southern blots and, in addition, requires significantly less DNA to perform these studies.

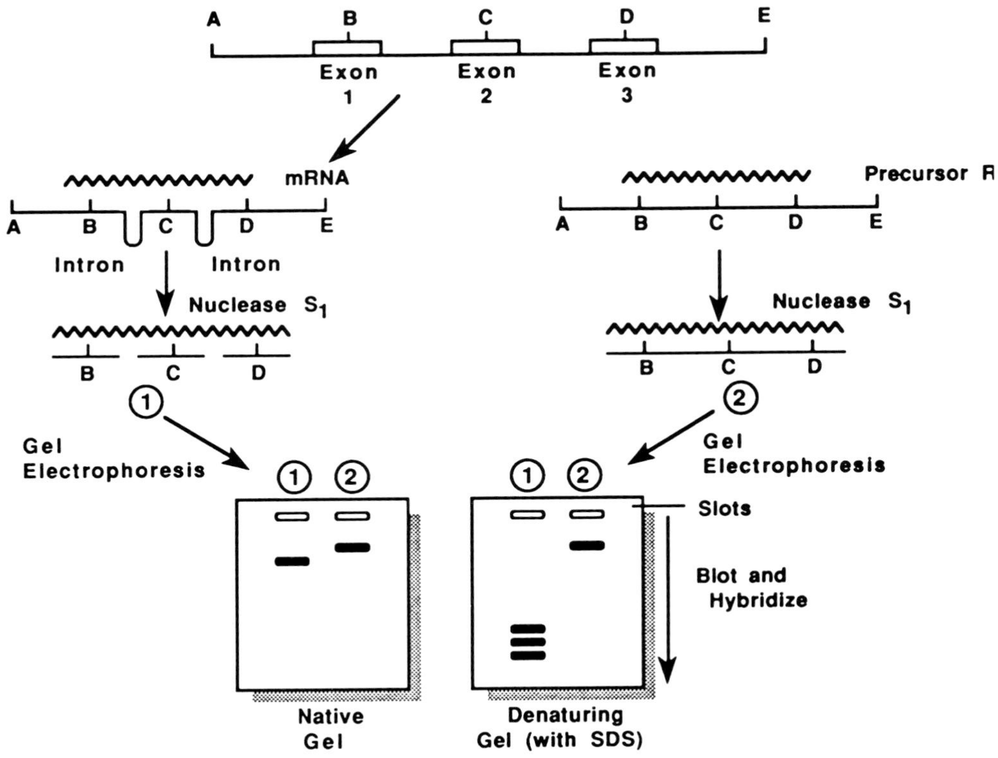

Figure 3.26 S1 nuclease protection analysis on Southern blots. The genomic DNA (*A* to *E*) is hybridized to an RNA population containing processed mRNA and precursor RNA. The RNA-DNA hybrids are treated with S1 nuclease [*(1)* and *(2)*] and analyzed on agarose gels. *SDS*, sodium dodecyl sulfate.

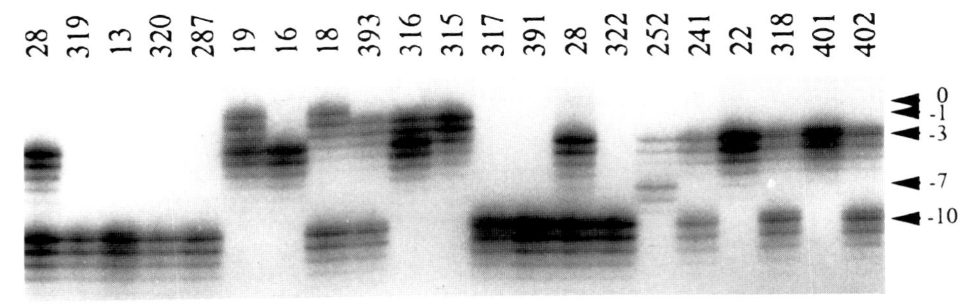

Figure 3.27 Sequence tagged sites (STS). This PCR-based method utilizes primers designed to amplify dinucleotide repeat sequences in various regions of the genome. The multiplicity of bands produces a VNTR, allowing for haplotype analysis and linkage analysis from small quantities of DNA. In this case, a primer pair designed to amplify a region of the X chromosome within intron 44 of the Duchenne muscular dystrophy or dystrophin locus amplifies a series of bands differing by 2 bp. The numbers at the top designate patient pedigree numbers.

Allele-Specific Oligonucleotide Analysis and PCR

The final method to be described is that of allele-specific oligonucleotide (ASO) analysis of amplified DNA by PCR. This method, described by Saiki et al. (75), is based on hybridization of a probe to amplified material. Here, a synthetic ASO probe, which is usually 19 to 20 bp long, is used to analyze amplified DNA that was spotted on a solid support such as nylon or nitrocellulose filters. When proper reaction conditions are used,

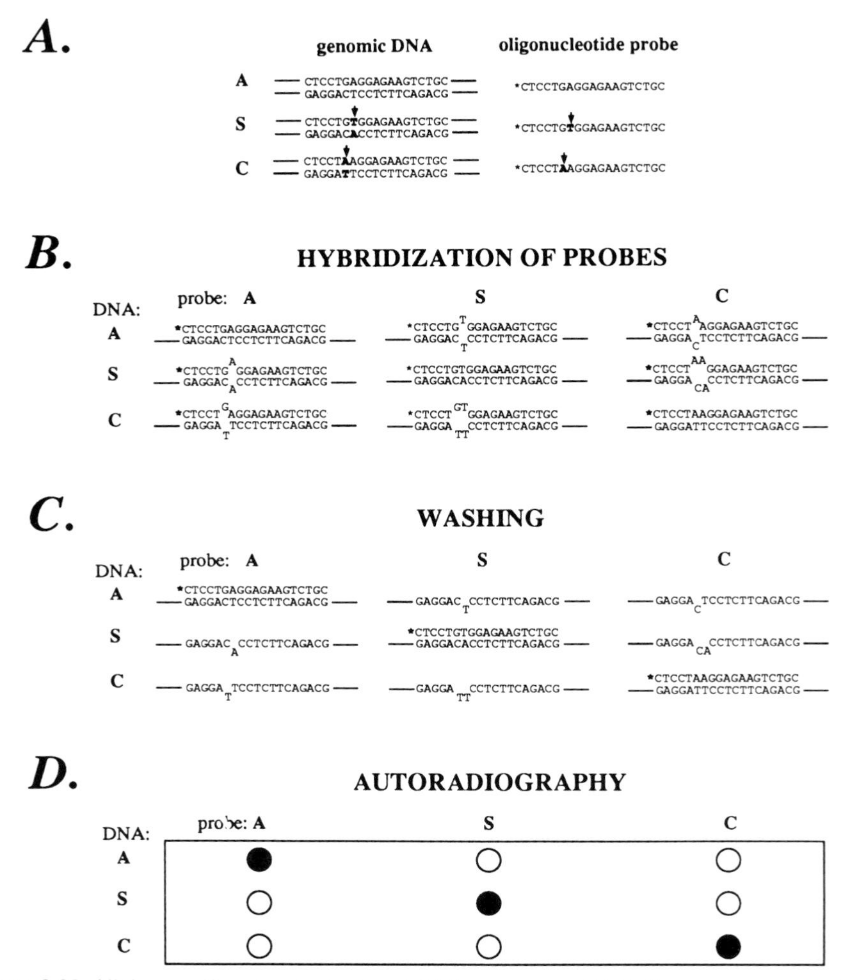

Figure 3.28 Allele-specific oligonucleotide (ASO) screening. (A) DNA sequence around the mutated site on the gene of interest. In this case, the sickle cell anemia mutation site in the β-hemoglobin gene is shown; allele A is the wild type and the mutations in alleles S and C are indicated with *arrows*. Each oligonucleotide probe is end labeled as indicated by the *asterisk*. (B) Using moderately stringent conditions, probes hybridize to all alleles, but where incomplete homology exists, one or two nucleotides mismatch. (C) At higher stringency of washing, only probes matching perfectly remain bound to genomic DNA. (D) Autoradiography detects the presence of bound, radiolabeled oligonucleotides and indicates the appropriate allele. *Reprinted by permission from Rossiter BJF, Caskey CT. FASEB J 1991;5:21–27.*

the ASO anneals only to perfectly matched sequences; single mismatches are sufficient to prevent hybridization. Using high stringency conditions to eliminate any potential for hybridization of partially mismatched probes to the target material, analysis is performed. The signal remaining at the end of the washing steps fully reflects the complementarity of the sequences of the amplified material. Using PCR, this in vitro amplification method produces a 10^5-fold increase in the amount of target sequence and permits analysis of allelic variation with as little as 1 ng of genomic DNA on dot blots or on Southern blots (Fig. 3.28).

Summary

This chapter has covered the majority of common molecular genetic technology, including the more advanced megabase methods. The methods described were selected because of their increased usage in cardiovascular molecular genetics and the increasing number of reports appearing in the general cardiovascular literature. It was the aim of this chapter to provide the background information that allows the reader to gain a new understanding of the increasingly common molecular genetic approaches to the cardiovascular system.

References

1. Blin N, Stafford DW: A general method for isolation of high molecular weight DNA from eukaryotes. *Nucleic Acids Res* 1976;3:2303.
2. Linn S, Arber W: Host specificity of DNA reduced by *Escherichia coli*. X: In vitro restriction of phage fd replicative form. *Proc Natl Acad Sci USA* 1986;59:1300.
3. Smith HO, Wilcox KW: A restriction enzyme from *Haemophilus influenzae*. I: Purification and general properties. *J Mol Biol* 1970;51:379.
4. Smith HO, Nathans D: A suggested nomenclature for bacterial host modification and restriction systems and their enzyme. *J Mol Biol* 1973;81:419.
5. Southern EM: Detection of specific sequences among DNA fragments separated by gel electrophoresis. *J Mol Biol* 1975;98:503.
6. Alwine JC, Kekmp DJ, Stark GR: Method for detection of specific RNAs in agarose gels by transfer to diazobenzyloxymethyl-paper and hybridization with DNA probes. *Proc Natl Acad Sci USA* 1977;75:5350.
7. Miller G, Lisco H, Stitt D: Establishment of cell lines from normal adult human blood leukocytes by exposure to Epstein-Barr virus and neutralization by human sera with Epstein-Barr virus antibody. *Proc Soc Exp Biol Med* 1971;137:1459.
8. Sela M, Anfinsen CB, Harrington WF: The correlation of ribonuclease activity with specific aspects of tertiary structure. *Biochim Biophys Acta* 1957;26:502.
9. Favaloro J, Reisman R, Kamen R: Transcription maps of polyoma virus-specific RNA: Analysis by two-dimensional nuclease S_1 mapping. *Methods Enzymol* 1980;65:718.

10. Stallcup MR, Washington LD: Region-specific initiation of mouse mammary tumor virus RNA synthesis by endogenous RNA polymerase II in preparations of cell nuclei. *J Biol Chem* 1983;258:2803.
11. Birnboim HC: Rapid extraction of high molecular weight RNA from cultured cells and granulocytes for Northern analysis. *Nucleic Acids Res* 1988;16:1487.
12. Glisin V, Crkvenjakov R, Byus C: Ribonucleic acid isolated by cesium chloride centrifugation. *Biochemistry* 1974;13:2633.
13. Ullrich A, Shine J, Chirgwin J, et al: Rat insulin genes: Construction of plasmids containing the coding sequences. *Science* 1977;196:1313.
14. Chomczynski P, Sacchi N: Single-step method of RNA isolation by acid guanidinium th iocyanate-phenol-chloroform extraction. *Anal Biochem* 1987;162:156.
15. Doty P, Marmur J, Eigner J, Schildkraut C: Strand separation and specific recombination in deoxyribonucleic acids: Physical chemical studies. *Proc Natl Acad Sci USA* 1960;46:461–476.
16. Chargaff E: Structure and function of nucleic acids as cell constituents. *Fed Proc* 1951;10:654.
17. Kelly RB, Cozzarelli NR, Deutscher MP, et al: Enzymatic synthesis of deoxyribonucleic acids. XXXII. Replication of duplex deoxyneonucleic acid by polymerase at a single strand break. *J Biol Chem* 1970;245:39.
18. Rigby PWJ, Dieckmann M, Rhodes C, Berg P: Labeling deoxyribonucleic acid to high specific activity *in vitro* by nick translation with DNA polymerase I. *J Mol Biol* 1977;133:237.
19. Feinberg AP, Vogelstein B: A technique for radiolabeling DNA restriction endonuclease fragments to high specific activity. *Anal Biochem* 1983;132:6.
20. Berkner KL, Folk WR: Polynucleotide kinase exchange reaction. Quantitative assay for restriction endonuclease-generated 5′-phosphoryl termini in DNAs. *J Biol Chem* 1977;252:3176.
21. Harrison B, Zimmerman SB: T4 polynucleotide kinase: Macromolecular crowding increases the efficiency of reaction at DNA termini. *Anal Biochem* 1986;158:307.
22. Church GM, Gilbert W: Genomic sequencing. *Proc Natl Acad Sci USA* 1984;81:1991.
23. Bowles NE, Olsen EGJ, Richardson PJ, Archard LC: Detection of coxsackie B-virus-specific RNA sequences in myocardial biopsy samples from patients with myocarditis and dilated cardiomyopathy. *Nature* 1982;278:416.
24. Grunstein M, Hogness DS: Colony hybridization: A method for the isolation of cloned DNAs that contain a specific gene. *Proc Natl Acad Sci USA* 1975;72:3961.
25. Benton WD, Davis RW: Screening λgt recombinant clones by hybridization to single plaques *in situ*. *Science* 1977;196:180.
26. Maxam AM, Gilbert W: A new method of sequencing DNA. *Proc Natl Acad Sci USA* 1977;74:560–564.
27. Sanger F, Nicklen S, Coulson AR: DNA sequencing with chain-terminating inhibitors. *Proc Natl Acad Sci USA* 1977;74:5463.
28. Gait MJ, Sheppard RC: Rapid synthesis of oligodeoxyneonucleotides: A new solid-phase method. *Nucleic Acids Res* 1977;4:1135.
29. Shine J, Seeburg PH, Martial JA, et al: Construction and analysis of recombinant DNA of human chorionic somatommamotropin. *Nature* 1977;270:494.

30. Mullis KB, Faloona FA: Specific synthesis of DNA *in vitro* via a polymerase catalyzed chain reaction. *Methods Enzymol* 1987;144:335.
31. Saiki RK, Scharf S, Faloona F, et al: Enzymatic amplification of β-globin genomic sequences and restriction site analyhsis of diagnosis of sickle cell anemia. *Science* 1985;230:1350–1354.
32. Oste C: Polymerase chain reaction. *Biotechniques* 1988;6:162.
33. Saiki RK, Gelfand DH, Stoffel S, et al: Primer-directed enzymatic amplification of DNA with a thermostable DNA polymerase. *Science* 1988;239:487–491.
34. Erlich HA, Gelfand D, Sninsky JJ: Recent advances in the polymerase chain reaction. *Science* 1991;252:1643.
35. Scharf SJ, Horn GT, Erlich HA: Direct cloning and sequence analysis of enzymatically amplified genomic sequences. *Science* 1986;233:1076–1078.
36. Frohman MA, Dush MK, Martin GR: Rapid production of full-length cDNAs from rare transcripts: Amplification using a single gene-specific oligonucleotide primer. *Proc Natl Acad Sci USA* 1988;85:8998–9000.
37. Lee CC, Wu X, Gibbs RA, et al: Generation of cDNA probes directed by amino acid sequence: Cloning of urate oxidase. *Science* 1988;239:1288.
38. Ochman H, Ajioka JW, Garza D, Hartl DL: Inverse polymerase chain reaction. In HA Erlich (ed): *PCR Technology. Principles and Application for DNA Amplification.* New York: Stockton Press, 1989, pp 105–112.
39. Nelson DL, Ledbetter SA, Corbo L, et al: Alu polymerase chain reaction: A method for rapid isolation of human specific sequences from complex DNA sources. *Proc Natl Acad Sci USA* 1989;86:6686.
40. Wrischnik LA, Higuchi RG, Stoneking M, et al: Length mutations in human mitochondrial DNA: Direct sequencing of enzymatically amplified DNA. *Nucleic Acids Res* 1987;15:529.
41. McMahon G, Davis E, Wogen GN: Characterization of C-K_1-ras oncogene alleles by direct sequencing and enzymatically amplified DNA from carcinogen-induced tumors. *Proc Natl Acad Sci USA* 1987;84:4974.
42. Gyllensten UB, Erlich HA: Generation of single-stranded DNA by polymerase chain reaction and its application to direct sequencing of the HLA-DQA locus. *Proc Natl Acad Sci USA* 1988;85:7652.
43. Orkin SH: Reverse genetics and human disease. *Cell* 1986;47:845.
44. Koenig M, Hoffman EP, Bertelson CJ, et al: Complete cloning of the Duchenne muscular dystrophy (DMD) cDNA and preliminary genomic organization of the DMD gene in normal and affected individuals. *Cell* 1987;50:509.
45. Kerem B-S, Rummens JM, Buchanan JA: Identification of the cystic fibrosis gene: Genetic analysis. *Science* 1989;245:1073.
46. Riordan JR, Rommens JM, Kerem B-S, et al: Identification of the cystic fibrosis gene: Cloning and characterization of complementary DNA. *Science* 1989;245:1066.
47. Lee WH, Booksteen R, Hong F, et al: Human retinoblastoma susceptibility gene: Cloning, identification and sequencing. *Science* 1987;235:1394.
48. Barker D, Wright E, Nguyen K, et al: Gene for Von Recklinghausen neurofibromatosis is in the pericentromeric region of chromosome 17. *Science* 1987;236:1100.
49. Gusella JF: DNA polymorphism and human disease. *Annu Rev Biochem* 1986;55:831.

50. Botstein D, White R, Skolnick M, Davis RW: Construction of a genetic linkage map in man using restriction fragment length polymorphisms. *Am J Hum Genet* 1980;32:314.
51. Ott J: *Analysis of Human Genetic Linkage*. Baltimore: Johns Hopkins University Press, 1985.
52. Brown WT, Wu Y, Gros AC, et al: RFLP for linkage analysis of fragile X syndrome. *Lancet* 1987;1:280.
53. Towbin JA, Brink P, Gelb B, et al: X-linked cardiomyopathy: Molecular genetic evidence of link to the Duchenne muscular dystrophy locus. *Pediatr Res* 1990;27:25A.
54. Bolhuis PA, Hansels GW, Hulsebus TJM, et al: Mapping of the locus for X-linked cardioskeletal myopathy with neutropenia and abnormal mitochondria (Barth syndrome) to Xz 28. *Am J Hum Genet* 1991;48:481.
55. Jarcho JA, McKenna W, Pare JAP, et al: Mapping a gene for familial hypertrophic cardiomyopathy to chromosome 14q1. *N Engl J Med* 1989;321:1372–1378.
56. Hejtmancik JF, Brink PA, Towbin J, et al: Localization of the gene for familial hypertrophic cardiomyopathy to chromosome 14q1 in a diverse American population. *Circulation* 1991;83:1592–1597.
57. Keating M, Atkinson D, Dunn C, et al: Linkage of a cardiac arrhythmia, the long QT syndrome, and the Harvey *ras*-1 Gene. *Science* 1991;252:704–706.
58. Kainulainen K, Pulkkinen L, Savolainen A, et al: Location on chromosome 15 of the gene defect causing Marfan syndrome. *N Engl J Med* 1990;323:935.
59. Kan YW, Dozy AM: Polymorphism of DNA sequence adjacent to human β-globin structural gene: Relationship to sickle mutation. *Proc Natl Acad Sci USA* 1978;75:5631.
60. Jeffreys AJ: DNA sequence variants in the γ, α and β-globin genes of man. *Cell* 1979;18:1.
61. Nakamura Y, Leppert M, O'Connell P, et al: Variable number of tandem repeat (VNTR) markers for human gene mapping. *Science* 1987;235:1616.
62. Gall JG, Pardue ML: Formation and detection of RNA-DNA hybrid molecules in cytological preparations. *Proc Natl Acad Sci USA* 1969;63:378.
63. Bender W, Spierer P, Hogness D: Gene isolation by chromosomal walking. *J Supramol Struct* 1979;10:32.
64. Poutska A, Pohl TM, Barlow DP, et al: Construction and use of human chromosome jumping libraries from NotI-digested DNA. *Nature* 1987;325:353.
65. Collins FS, Drumm ML, Cole JL, et al: Construction of a general human chromosome jumping library with application of cystic fibrosis. *Science* 1987;235:1046.
66. Schwarz DC, Cantor CR: Separation of yeast chromosome-sized DNAs by pulsed field gradient gel electrophoresis. *Cell* 1984;37:67.
67. Carle GF, Olson MV: Separation of chromosomal DNA molecules from yeast by orthogonal field alteration gel electrophoresis. *Nucleic Acids Res* 1984;12:5647.
68. Burke DT, Carl GF, Olson MV: Cloning of large segments of exogenous DNA into yeast by means of artificial chromosome vectors. *Science* 1987;236:806.
69. Shows TB, Sakaguchi AY, Naylor SL: Mapping the human genome, cloned genes, DNA polymorphisms and inherited disease. *Adv Hum Genet* 1982;12:341.

70. Cotton RGH, Rodrigues NR, Campbell RD: Reactivity of cytosine and thymine in single base pair mismatches with hydroxylamine and osmium tetroxide and its application to the study of mutations. *Proc Natl Acad Sci USA* 1988;85:4397–4401.
71. Davidson JN: *The Biochemistry of the Nucleic Acids*, 7th ed. New York: Academic Press, 1972.
72. Myers RN, Larin Z, Maniatis T: Detection of single base substitutions by ribonuclease cleavage at mismatches in RNA:DNA duplexes. *Science* 1985;230:1242.
73. Winter E, Yamamoto F, Almoguera C, Perucho M: A method to detect and characterize point mutations in transcribed genes: Amplification and overexpression of the mutant c-K_1-*ras* allele in human tumor cells. *Proc Natl Acad Sci USA* 1985;82:7575.
74. Berk AJ, Sharp PA: Sizing and mapping of early adenovirus mRNAs by gel electrophores of S_1 endonuclease digested hybrids. *Cell* 1977;12:721.
75. Saiki RK, Chang C-A, Levenson CH, et al: Diagnosis of sickle cell anemia and β-thalassemia with enzymatically amplified DNA and nonradioactive allele-specific oligonucleotide probe. *N Engl J Med* 1988;319:537.

Chapter **4**

The Essentials of Molecular Genetics

Nature is nowhere accustomed more openly to display her secret mysteries than in cases where she shows traces of her workings apart from the beaten path; nor is there any better way to advance the proper practice of medicine than to give our minds to the discovery of the usual law of Nature by careful investigation of cases of rare forms of diseases. For it has been found that in almost all things, that what they contain of useful or applicable nature is hardly perceived unless we are deprived of them, or they become deranged in some way.

William Harvey, 1657

There are many disorders of the myocardium that are produced by inherited defects. Throughout the ages the history of medicine is highlighted with important observations learned from nature's mistakes that have provided insight fundamental to our understanding of the etiology of disease. We have not had the same opportunities in cardiology. However, this is likely to change. The first inherited primary cardiomyopathy—hypertrophic cardiomyopathy (HCM)—has succumbed and has been mapped to chromosome 14. The cardiomyopathies have been very neglected relative to ischemic heart diseases. Unravelling the molecular basis for these diseases should provide, we hope, for better diagnosis and treatment and ultimately prevention of the disease. In addition, diseases whose primary effect is on the myocyte have the advantage of providing information that is likely to be fundamental to our understanding of compensatory cardiac growth, as opposed to ischemic heart disease, which is a disease of the coronary arterial wall with its effect on the myocardium being secondary. It is estimated that there are more than 70 inherited diseases that affect the heart. The first gene to be isolated for a disease affecting primarily the myocardium was that for HCM. The locus on which the gene resides was mapped to chromosome 14q1 in 1989 and the gene was identified in 1990 as β-myosin heavy chain. Since that time the chromosomal loci for the genes responsible for two other cardiac diseases have been mapped, and the rate of locus mapping is likely to accelerate at a very rapid pace. Several other diseases, such as Duchenne muscular dystrophy, which is associated with a myopathy of skeletal and cardiac muscle, have been mapped. The expanding number of available DNA markers and the widespread application of the technique of polymerase chain reaction (PCR) will significantly hasten our ability to localize and isolate disease-related genes. There are now over 3,000 DNA markers for which we know the chromosomal locations, and over 500 of these are disease-related genes. While cell culture and animal models are essential to our attempts to understand human disease, the ultimate understanding is likely to come from a molecular analysis of the disease in its natural habitat, humans. Actual isolation of a mutated gene and its altered protein for which the functional, structural, and clinical manifestations are known represents the ultimate experiment.

The developments in molecular biology that make it more feasible to isolate genes today than in the past are many, but major among these advancements have been computerized linkage analysis and the technique of detecting DNA polymorphism by restriction endonucleases, referred to as *restriction fragment length polymorphism* (RFLP). Full exploitation of the potential of linkage analysis requires the clinician to play a pivotal role in isolating the genes responsible for hereditary diseases. First, chromosomal mapping by linkage analysis requires access to family pedigrees that, to be useful, must undergo clinical assessment and definitive diagnosis. Interpretation of the patient's DNA analysis of RFLP and the results of the linkage studies will be misleading if the diagnosis is incorrect. Second, the clinician will be responsible for obtaining cardiac tissue as well as performing the physiological studies necessary to show the causal link between the defective gene and the recognized phenotype. The application of molecular genetics demands that both the investigator at the molecular level and the clinician investigating the families work as a team. Otherwise the defects responsible for the disorders will remain unknown and we will miss the opportunity for specific therapy, including that of gene replacement.

Positional Cloning (Reverse Genetics)

Knowledge of the protein abnormalities of a hereditary disorder significantly increases one's chances to identify and isolate the causative gene. This is possible through determining the sequence of the proteins involved and deducing their expressed nucleotides sequences (complementary DNA sequences). Probes can then be designed to identify the approximate chromosomal location and ultimately clone and sequence the gene and the specific mutation. Familial hypercholesterolemia and some of the thalassemias are disorders whose genes were isolated and cloned using this classic approach. Unfortunately, for the majority of the diseases, we do not know the defect or protein, so this approach cannot be used. However, as a result of the recent advances in linkage analysis and the development of DNA markers detected as RFLPs, it is now possible to isolate the genes in these diseases (1,2). This approach, known until recently as reverse genetics, is more correctly termed *positional cloning* since it involves cloning of a disease-related gene from knowing only its chromosomal position. The initial process involves mapping the gene to its particular chromosomal region by linkage analysis. Since the human genome is more than 3 billion base pairs (bp), this is a very tedious, cumbersome, and slow process. It is made possible by the linking of a disease-related gene to a marker of known chromosomal position (locus) of the human genome identified by RFLP as discussed in the section on "Isolation and Identification of the Gene" (3). Once a disease is linked to a marker of known chromosomal locus, it follows that the disease locus is on the same chromosome and its approximate position on the chromosomal is known. One then attempts to identify other DNA markers that flank the disease locus in as close proximity as possible. Having established closely flanking DNA markers, it is then possible by the technique referred to as chromosomal walking to isolate and clone the intervening region containing the gene of interest. One would like to identify the gene and ultimately sequence it to determine the precise mutation causing the disease. The remaining task, if not already apparent, would be to determine the gene product (protein) and the pathophysiology of how it induces the disease.

The overall approach to chromosomal mapping of hereditary diseases by linkage analysis and subsequent isolation of the gene may be summarized categorically as follows: 1) collection of data from families having individuals affected by the specific disease through two or three generations; 2) clinical assessment to provide an accurate diagnosis of the disease using consistent and objective criteria to separate normal individuals from those affected and those who are indeterminate or unknown; 3) collection of blood samples for immediate DNA analysis and to develop lymphoblastoid cell lines for a renewable source of DNA; 4) development of a pedigree analysis of the family; 5) analysis of the DNA for RFLPs using initially a large number of DNA markers of known chromosomal loci that span the human genome in an attempt to find a known locus that is linked to the disease; 6) development of flanking markers around the region containing the disease locus; 7) isolation and cloning of the region of DNA containing the gene; 8) identification of the gene; 9) sequence analysis of the gene to identify the precise mutation(s) causing the disease; 10) demonstration of a causal relationship between the defective protein and the disease; and 11) development of a convenient test to screen for the mutations.

DNA is a monotonous, repetitive molecule consisting of just four different nucleotides. Despite its repetitive nature, there are the discrete DNA units that code for polypeptides referred to as genes. Isolation of genes responsible for disease is extremely difficult when one does not know the protein product. Part of the difficulty in determining the location of a gene is the lack of known landmarks along the DNA molecule. The chromosome consists of one continuous DNA molecule. The smallest chromosome, 21, has 50,000 bp and chromosome 1, the largest, has 250 million bp. Determining where on a chromosome a particular gene is located, referred to as the *locus* of the gene, is termed *chromosomal mapping.* Chromosomal mapping of those diseases for which the defective protein is not known is done by linkage analysis, which makes use of modern molecular techniques and computerized genetic statistics. In brief, this consists of showing that the gene of interest is linked to a DNA marker of known chromosomal locus. It is possible on the basis of linkage analysis alone to construct a chromosomal map in which the distance between the various markers is estimated in units referred to as centimorgans (cM), named after Morgan, the great modern geneticist. A centimorgan is approximately 1 million bp; however, it is important to emphasize that this is a statistically derived genetic map and the distances are only approximate. The desired chromosomal map is a physical map in which the distance is determined in base pairs as a result of cloning and sequencing the gene and, thus, represents the actual distance rather than an estimate. However, in those diseases in which the protein product is not known, the approach is generally to first determine the chromosomal locus from a genetic map based on linkage analysis, followed by development of the physical map after isolation and sequencing of the specific gene. To appreciate the ingenious nature of linkage analysis and the basis for its complexity, certain background information is essential, which is discussed subsequently. Mapping the chromosomal location of genes by linkage analysis requires knowledge of how one recognizes DNA markers and an understanding of the concept of linkage.

Concept of Linkage

There are well over 3,000 regions of DNA that can be recognized by appropriate DNA markers (4,5). In brief, the objective of linkage is to show that the gene of interest is linked to one of these markers, which means both loci are on the same chromosome and in relatively close physical proximity. Given that there are 3 billion bp and 3,000 markers, there should be a marker every 1 million bp; thus, the chances of linking a gene of interest to one of these markers should be rather good. While it is now conceivable that, with the necessary resources and expertise, one could map practically any disease, the process may be very tedious and laborious. Despite there being a marker on the average of every 1 million bp, they are not evenly distributed, and some chromosomes have some areas of up to 50 million bp without a known marker. Nevertheless, the object of linkage analysis is to try to link the gene of interest to one of the markers of known chromosomal location, from which can be derived the chromosome on which the gene resides and the gene's approximate location. This is done through linking the disease-related locus to that of a known chromosomal marker, which is cone by showing that the two loci are coinherited in various members of a family more commonly than by chance alone. Genetic loci are said to be linked if they are coinherited within a family. This coinheritance is due to the physical proximity of the marker and the disease loci on the chromosome. An ideal disease for linkage analysis is one that follows Mendelian inheritance and in which the mutation(s) are all in a single gene. It is possible to account for genetic heterogeneity, and of course we know that many diseases are due to multiple genes (referred to as *polygenic*) and other diseases are *multifactorial* in that they are the result of an interaction between environmental factors and many genes. It is possible, and significant efforts are now being undertaken with the combination of linkage analysis and sophisticated molecular techniques, to unravel diseases of polygenic origin as well as those of multifactorial origin. However, for illustrative purposes our discussion focuses on those diseases that are due to a single gene, of which we know there are many affecting the heart and other organs.

Since humans inherit two sets of chromosomes (diploid), all of the genes carried by the autosomal chromosomes have two forms, referred to as *alleles*. The two alleles occupy the same chromosomal locus but the loci are on different chromosomes (homologous), which gives rise to the terminology of homologous loci on homologous chromosomes. One allele comes from the mother and the other from the father. In a Mendelian dominant disease, the individual with the disease will have one defective allele that is responsible for the disease and one normal allele. Why the defective allele is expressed and not the normal allele remains unknown. In some individuals, despite having the disease-related gene, there is no clinical evidence of the disease. Thus the genotype is that of the disease but the phenotype is normal. The percentage of individuals with the disease-related gene who do have some form or features of the disease is referred to as the *penetrance* of the disease. This is to be distinguished from *expressivity*, which refers to the variable nature of the clinical features. Penetrance is an all-or-none phenomenon, meaning any manifestation of the disease is considered penetrance, whereas expressivity refers to the variability of the clinical findings (phenotype) in those individuals in whom the disease is present.

The concept of linkage analysis is illustrated in Figure 4.1. Shown in the right-hand panel is an illustration of genetic linkage between a locus for a DNA marker and that of a disease that is inherited in a Mendelian dominant fashion. The locus designated with an "A" carries the allele responsible for the disease. The corresponding locus "a" on the homologous chromosome has the allele that codes for the same protein but has not undergone a mutation and is thus the normal allele. The loci

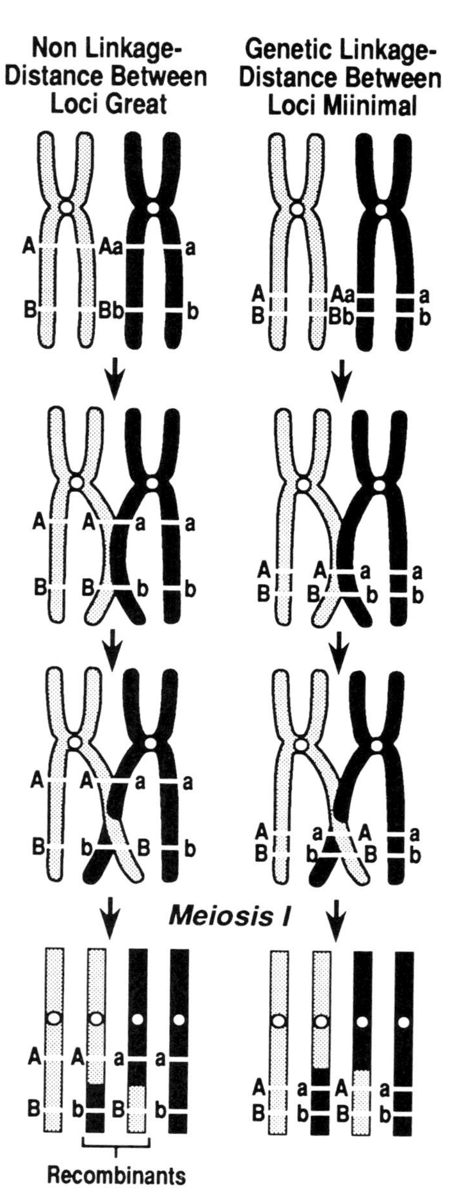

Figure 4.1 The differences between two genetically linked loci (*right-hand panel*) and two loci that are not linked (*left-hand panel*). In the left-hand panel, the loci designated "A" represent the locations of two homologous genes and those designated "B" are separate loci for two homologous genes. The other pair of homologous chromosomes with loci designated "a" and "b" represent the homologous genes from the other parent. Loci "A" and "B," as are "a" and "b," are very far apart and, upon the formation of a chiasma, the chromosome exchanges identical portions, with the result that the portion containing the "B" locus now recombines with the "a" locus. In genetics parlance, recombination has occurred whereby the two loci "A" and "B" are now on separate chromosomes, as are "a" and "b." In contrast, in the right-hand panel the loci "A" and "B" and "a" and "b" are in close physical proximity. During the formation of a chiasma and recombination, because the two loci are so close together they both move to the other chromosome. Thus, "A" and "B," as are "a" and "b," are genetically linked: the "A" and "B" loci will stay together more frequently than by chance alone, meaning in more than 50% of the meioses. It should not be construed, however, that if two markers are genetically linked recombination never occurs. Because the two markers are very close together (within a few hundred kilobase pairs), they may never separate, whereas at a distance of 20,000 kbp they would be expected to separate with a frequency of 20% in subsequent meioses.

designated "B" and "b" represent alleles of a DNA marker of known location that has nothing to do with the disease. In the right-hand panel the disease and the marker loci are so close that they tend to be coinherited with the family, whereas in the left-hand panel the DNA marker of known location is so far from the locus carrying the disease allele that it is far less likely they will be coinherited. Prior to meiosis, homologous chromosomes, and only homologous chromosomes, come together and form bridges (chiasmata) between them such that segments of equal proportion are exchanged between them, giving rise to crossover of various genes. This separates various loci with their genes that otherwise would have been coinherited. This is the basis for genetic diversity within the species. However, it is important to point out that this process, even with crossover, makes certain that the allele crossing over to its homologous chromosome will occupy the same location (locus) as it did on its previous chromosome. Thus, although two homologous alleles can exchange positions between the two homologous chromosomes, the actual position on either chromosome, referred to as the *locus*, remains the same for any particular allele.

In genetics parlance, crossing over is referred to as *recombination*; that is, a segment of one chromosome has broken away and recombined with its homologous chromosome. The recombination frequency or fraction (θ) between two markers is the ratio of the number of crossover events to the total number of meioses. The lower the recombination frequency between the locus of a marker and that of disease-related gene, the closer those two must be in physical distance on the chromosome. Even though the locus of the marker and that of the disease-related gene are in close enough proximity to be genetically linked, recombination may occur, and the extent to which recombination does occur reflects roughly the physical distance between the two loci. Theoretically, if two loci are coinherited more than 50% of the time they are said to be genetically linked. There is a 50% chance of coinheritance of two loci even if they are not linked or even if they are on different chromosomes. However, genetic linkage means they are on the same chromosome and in reasonably close physical proximity. The recombination fraction is used to develop a means of estimating the genetic distance between the linked loci. The genetic distance is measured in centimorgans, with 1 cM equivalent to a 1% recombination frequency. It has also been determined that 1 cM approximates very roughly 1 million bp in physical distance. This correlation is variable from chromosome to chromosome and from region to region even on the same chromosome; for example, recombination is more frequent in the telomeric than in the centromeric portion of the chromosome. It should be realized that if the marker locus and the disease-related locus are close (e.g., within 5 to 10 cM) then a single crossover may be uncommon and a double crossover rare. However, two loci may be 20 to 40 cM apart and a double crossover can occur, which recombines the locus with the original chromosome, leading to coinheritance of the two (linkage of the two loci). When this occurs, the genetic distance is misleading and represents a gross underestimation of the true physical distance between the two loci.

Axioms Fundamental to Linkage Analysis

The concept of linkage analysis and an appreciation of how to apply it in analyzing Southern blotting of DNA is often difficult in part because certain axioms are not emphasized. If these axioms are not properly understood, it probably means that the basis for linkage analysis is not understood. The axioms in applying linkage analysis to a dominantly inherited disease are as follows:

1. In reference to genes on autosomal chromosomes, every individual has two forms of the gene, referred to as alleles, one being inherited from the mother and the other from the father. In individuals with dominant disease, one allele is defective and the other is normal.
2. The DNA marker of known chromosomal location to which a disease gene is linked also has two alleles, one from the father and one from the mother.
3. When a DNA marker and a disease-related gene are said to be genetically linked, it means that the two loci are linked, not their alleles.
4. Since the two loci are linked, and not the alleles, which allele a particular offspring gets is total chance since the inheritance of either or both alleles at a particular locus is independent of the other.
5. Neither of the alleles at the marker locus has anything to do with causing the disease. Both alleles of the marker locus occur in the general population and do not themselves cause disease. They simply reside at a locus that is in close enough physical proximity to the locus that contains the disease-producing gene to be coinherited more often than by chance.
6. To be informative for linkage analysis, the alleles of the marker locus and the disease-related locus must be heterozygous. This means that the two alleles at the marker locus must not have an nucleotide sequence identical to that of the probe being utilized for their detection but must be polymorphic; similarly, for the disease locus one allele must be normal and the other abnormal.
7. Linkage of a marker locus and a disease-related locus implies that the two are coinherited more often than by chance alone, which means more often than 50% of the time. It does not mean, however, that the two loci are always coinherited; in fact, only if they are extremely close would this be true.
8. It follows from previous axioms that, in analyzing the DNA of the marker loci of individuals within families affected with the disease, the same pattern may be seen in an individual without the disease as in those individuals with the disease. This is why computer analysis is necessary to ascertain whether the disease allele is more commonly inherited with one or more of the alleles at the marker loci than would be expected by chance.
9. Crossover or recombination occurs between, and only between, homologous chromosomes, so the alleles that cross over or recombine occupy the same locus on their new chromosomes as they did on the previous ones.

It is evident from these axioms that the linkage is not necessarily between the defective allele causing the disease and any one specific allele at the chromosomal marker locus. While the alleles of two genetically linked markers tend to be coinherited, either allele (or both in equal proportions) can occur with the disease allele, even within a single family. Thus, while one marker allele may occur in most members of a single family, members of that same family who do not have the defective allele for the disease may have the same polymorphic pattern at the marker locus. The corollary of this is also true; or other alleles(s) of the marker may also occur in the diseased members of the family, since the two loci are linked but inheritance of alleles at both loci occurs independently. In addition, genetic linkage in a family pedigree does not imply that the diseased allele occurs more often in the general population with a specific allele at the marker locus. Thus, the diseased allele, either in a particular family or in the general population, can be coinherited with any of the alleles at the marker locus. These features are illustrated in Figures 4.2 and 4.3.

Development of a Genetic Map from Linkage Analysis

The recombination frequency (θ) between two markers is the percentage of the total number of meioses in which a detectable crossover event occurs between them. When two markers are close together, the likelihood of crossover between them is small and the likelihood of more than one crossover between them is negligible, so that θ is approximately equal to the genetic distance. It should be noted that if two (or any even number of) crossover events occur between two loci, the markers will end up once more on the same chromatid and will appear not to have separated (recombined). Genetic distance is measured in centimorgans, and over small distances 1 cM equals approximately 1% recombination ($\theta = 0.01$). Over greater distances, double recombinants occur and decrease the recombination frequency relative to the genetic distance. On average, 1 cM equals about 1,000 kilobase pairs (kb), but this relationship is far from constant and varies for different chromosomes, different regions of the same chromosome, and the gender of the individual studied.

Genetic distance is calculated from the recombination frequency using various formulas. In doing so one assumes that an initial crossover has no effect on the likelihood of a second crossover event within that region, which is often referred to as *no interference*. One also assumes that there is no difference in the crossover rate, which is known not be be true; crossover frequency is much higher at the ends of the chromosomes (telomeres) than at the center of the chromosomes (centromeres). It is also known that there is a higher crossover rate in females than in males. The genetic distance, based on the recombination fraction,

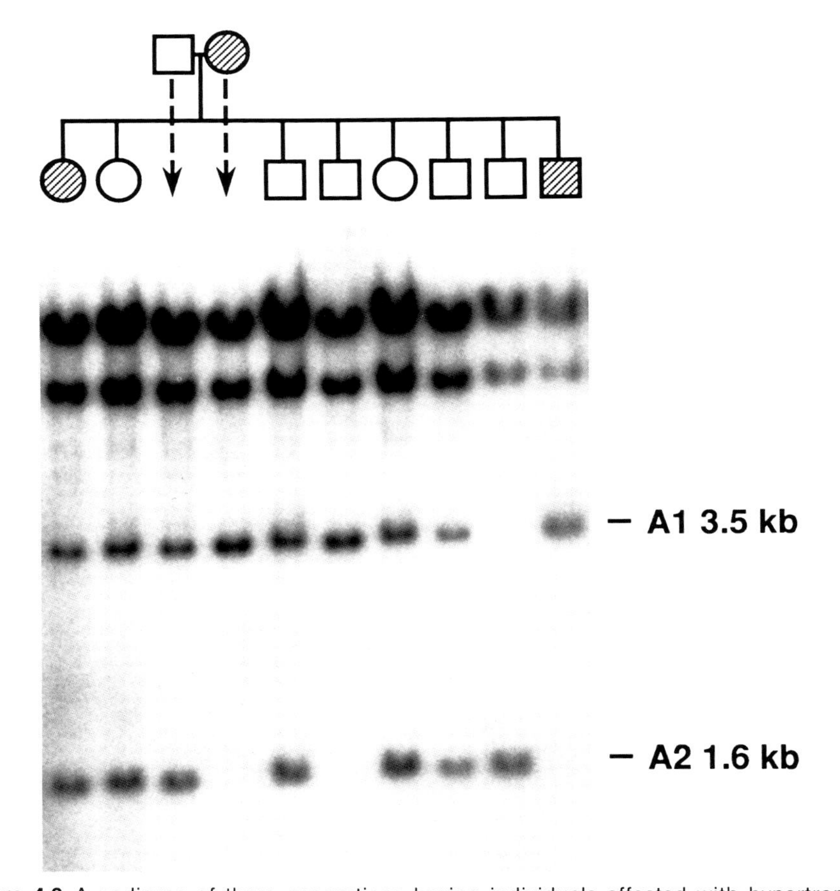

Figure 4.2 A pedigree of three generations having individuals affected with hypertrophic cardiomyopathy (HCM). The *open circles* indicate unaffected females, the *open squares* indicate unaffected males, the *solid symbols* indicate affected individuals (both male and female), the *slash* through a symbol indicates the patient is dead, and a circle or square within the circle or square indicates the diagnosis is uncertain. DNA was analyzed for restriction fragment length polymorphisms (RFLPs) by Southern blotting, and the results are shown on this autoradiograph for 11 of the individuals in the pedigree. Each vertical lane represents the DNA of the individual indicated by the number above, which corresponds to the same number on the pedigree. The DNA was digested with the restriction endonuclease *Taq*I and separated on agarose gel electrophoresis. It was then denatured into its two separate strands, transferred to a nylon membrane by the Southern transfer technique, and probed with a ^{32}P-labeled probe. The probe, referred to as P436, was derived from part of the β-myosin gene, which is known to be located on the long arm of chromosome 14. This probe recognizes two alleles, one at 4.2 kb and the other at 1.8 kb. As illustrated, the smaller fragment of 1.8 kb has migrated further than the larger fragment of 4.2 kb. The larger fragment at the top is consistently present in all of the individuals, so we will be examining the polymorphic alleles of 4.2 kb (A1) and 1.8 kb (A2).

Individual 51 is an affected female who is heterozygous, having received the A1 allele from one of her parents and the A2 allele from the other parent. Individual 49, in contrast, is homozygous, having inherited the identical A2 allele from both the mother and the father. Individual 53, a normal female, is also homozygous for the A2 allele. Individual 57, a normal female, is heterozygous at this locus, having both the A1 and A2 alleles. Individual 59, an affected male with HCM, is also heterozygous. Individual 64, an affected male, is homozygous, with both alleles being A2. Individual 66, an affected female, is heterozygous, having both the A1 and A2 alleles. Individuals 67 (normal male), 72 (normal female), and 78 (affected estimates a distance that approximates that of the physical distance. Second, unlike recombination fractions, genetic distances are additive, so the distance between two markers should equal the sum of the distance between all of the intervening markers. Finally, the concept of map distance formalizes the linear arrangement of genes into a genetic map, an extremely useful concept when combined with modern recombinant DNA technology.

Estimation of the Odds of Genetic Linkage

There are several methods to determine if two marker loci (or a disease and a marker) are linked, but the one used most commonly is that of the maximum likelihood estimation (6). One can estimate the probability of a particular inheritance pattern for a set of markers based on the linkage relationship between these markers. This probability can then be compared to the probability of that particular inheritance pattern appearing if the markers were not linked. The ratio of these two probabilities (i.e., of linkage at a given recombination fraction versus nonlinkage) is called the *odds ratio for linkage* at that recombination fraction. This ratio is usually expressed as a logarithm. The value is called the logarithm of the odds, or lod, score. Thus, a lod score of 1 represents 10^1:1 or 10:1 odds that a marker is linked. The lod score is usually calculated for a series of recombination fractions between two markers, and the recombination fraction giving the highest lod score is thus the relationship with the highest probability of being the true value, the maximum likelihood estimate of θ.

---

female) are all homozygous for the A2 allele. Individual 79 is a normal male and is heterozygous, having both A1 and A2 alleles. Computer analysis of the β-myosin gene in this family together with that in other families showed linkage between this marker and the disease for hypertrophic cardiomyopathy. A lod score was obtained of greater than 4, indicating the odds for linkage are more than 99%.

The analysis of this Southern blot illustrates several of the key features of linkage analysis explained in the text: 1) the same polymorphic pattern at the marker locus can be seen in both a normal and an affected individual within the same family, and 2) some affected individuals are homozygous at the marker locus while others are heterozygous, and, as indicated in the text, only those individuals who are heterozygous for the two alleles will provide information for linkage analysis with this particular marker locus. Thus, which allele is inherited by the sibling from the parents at the marker locus is completely random and independent of which allele is inherited at the disease gene locus, despite the two loci being linked. The analysis in this family also shows how, because of the lack of information, one may require a larger number of individuals than initially expected to ascertain whether linkage is present between the marker locus and that of the disease. In several of the individuals shown here the marker locus is homozygous and therefore will contribute almost no information to the linkage analysis. For a probe to be informative it must be heterozygous, which is frequently not the case, as illustrated in this pedigree analysis.

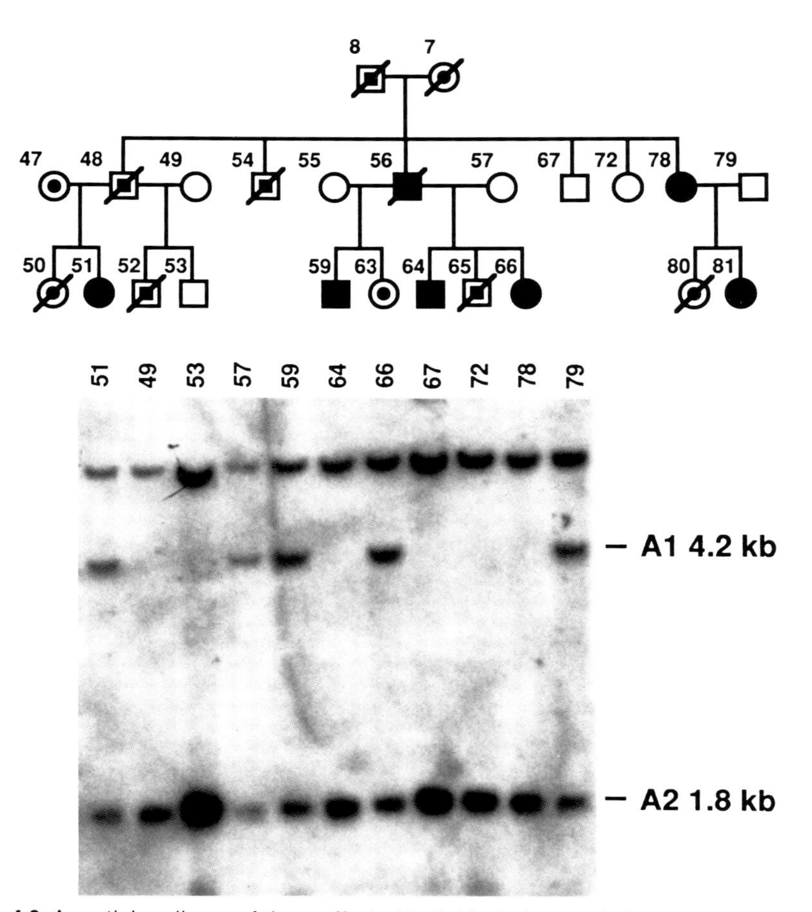

Figure 4.3 A partial pedigree of three affected individuals (*open circle or square*) and seven unaffected individuals (*shaded circle or square*). Below the pedigree is shown the results of their DNA analysis, after Southern blotting, using a myosin probe (PSC14) digested with the restriction endonuclease *BamI*. The features illustrated in this pedigree are similar to those shown in Figure 4.1. Each lane represents the DNA of the individual shown by the pedigree symbol above it. The two upper bands on the blot are constant in all the individuals, and the two alleles that show polymorphism are A1 and A2, indicated on the right as being 3.5 and 1.6 kb, respectively. Illustrated here are affected individuals who are both heterozygous for the two alleles, as indicated by the first affected female, and homozygous for the two alleles, as indicated by the remaining two affected individuals, one being homozygous for the A1 allele and the other being homozygous for the A2 allele. Similarly, some of the unaffected individuals are homozygous and others are heterozygous, and the patterns of the affected and unaffected individuals cannot be differentiated by pure visualization. Computer analysis is necessary to ascertain whether this particular locus is linked to the disease locus.

While lod scores are useful values, their interpretation is not straightforward. A lod score of 3 or greater is usually considered strong evidence of linkage. While a lod score of 3 represents 1,000:1 odds in favor of linkage by the results of the linkage analysis, it does not take into account the strong a priori odds against linkage. Since there are 46 chromosomes, given any two randomly selected loci, their chances of being on different chromosomes are about 50 times more likely than not, so the odds against linkage are 50:1. Thus, a lod score of 3 corresponds to 20:1 in favor of linkage, which means a lod score of 3 will prove spurious 1 time in 20 (95% likelihood of linkage), and a lod sore of 4, 1 time in 200 (99% likelihood of linkage). The a priori odds against linkage are considerably smaller if a disease is known to be inherited in an X-linked fashion. Thus, a lod score of 2 is considered significant evidence in favor of linkage for X-linked diseases. In similar fashion, a lod score of -2 or less is considered significant evidence against linkage for autosomal or X-linked diseases.

Identification of DNA Markers by RFLP

Each gene other than those that reside on the X chromosome is encoded by two different alleles. We inherit one allele from each parent, and they consistently occupy the same chromosomal locus on homologous chromosomes. It was recognized in the late 1970s that the DNA of one individual's genome, when compared to that of another individual, shows a difference in the sequence of its base pairs (polymorphism) every 300 to 500 bp. Polymorphisms occur more frequently in the sequence of the unexpressed DNA (introns) than in DNA coding for proteins (exons), especially if the latter polymorphisms result in a change in the protein sequence. This is because any sequence changes in proteins may interfere with their function and make individuals who carry them less fit.

Restriction endonucleases cleave foreign DNA by recognizing specific DNA sequences three to eight base pairs (bp) long. Consequently, digestion of human DNA by a given restriction endonuclease results in a specific pattern of fragments characterized by the number of fragments cleaved and the length of each fragment. These DNA fragments can be separated by size using agarose gel electrophoresis (Southern blotting) and detected by hybridization to labeled DNA probes. If there were a base change that altered the recognition site of a restriction endonuclease or created an additional recognition site in one allele but not in the other, the patterns exhibited by the DNA on gel electrophoresis after digestion by that specific enzyme would be distinctly different for the two alleles. Thus, digestion of DNA by a restriction endonuclease followed by separation of the fragments by electrophoresis and detection of the specific fragments by labeled DNA probe provides a means to detect the minor sequence differences, or restriction fragment length polymorphisms (RFLPs). It is of course important to realize that the DNA probe will detect the polymorphism only if it is lacking or if an extra restriction site is in the region of the DNA to which the probe will hybridize. These polymorphisms between two alleles of the same homologous loci can be localized to provide a landmark along the chromosomal DNA to which other genetic markers, including disease loci, can be linked. This is the basis for RFLP and how one uses Southern blotting to identify polymorphic markers of known chromosomal location (7,8) (see Chapter 3). An illustration of how the Southern blotting technique is used to identify RFLPs in a family pedigree is shown in Figures 4.2 and 4.3.

Isolation and Identification of the Gene

Linkage analysis provides a means to map the disease to a region of a specific chromosome; however, the position is only an approximation. It also provides an estimate of the distance between the disease-related locus and that of the marker(s). Since the resolution of linkage analysis is seldom better than about 1,000 kb, or 1 megabase pair (Mb), it means that the distance between the marker and the disease-related locus is at least this far in some cases but may be 5 Mb or more in others. A distance of even 1,000 kb is formidable using conventional cloning techniques. Until recently, it was only possible to clone DNA fragments up to 50 kb using cosmid vectors, thus making the task of cloning a segment of 1,000 kb nearly impossible. Therefore, one makes every attempt to develop closer markers flanking the region of interest before pursuing conventional cloning techniques in the hope of isolating the gene of interest.

The conventional approach is to clone overlapping DNA fragments from the region between the disease-related locus and that of the marker. This may be referred to as *restriction fragment mapping* (9), which consists of selecting restriction endonucleases that will cut the DNA at different, but known, specific sites to include ultimately all of the intervening region. These fragments resulting from digestion with the endonucleases are separated by electrophoresis and their electrophoretic patterns are compared for similarities. A map is constructed of these restriction fragments with the hope that there is enough overlap of the various fragments to provide one with the means to map the total distance. The process of aligning the digested fragments in their appropriate sequence is aided by knowing the sequences of the recognition sites used by the endonucleases for digestion.

Frequently, one must cover distances much larger than can be approached even through chromosome walking. Another approach is that of chromosomes jumping (10), a procedure whereby the DNA is cut by special endonucleases that provide large DNA segments. The ends of these segments are ligated together into a circle to bring portions of the DNA that were previously separated by considerable physical distance into close proximity. The circularized DNA is digested and the products are cloned into appropriate vectors. The clones containing the segment of interest are then identified using a DNA probe. The process is repeated sequentially until the desired distance is traversed and the gene isolated. Using this technique, for example, investigators were able to travel approximately 100 kb to the gene for cystic fibrosis and 200 kb along the region containing the gene for Huntington disease. However, recent development of techniques designed to analyze large DNA fragments (up to 2,000 kb) promise to significantly accelerate the rate of chromosome walking. These techniques consist of yeast artificial chromosome (YAC) cloning (11) and pulsed-field gel electrophoresis (PFGE) (12). YAC cloning provides the opportunity to clone fragments up to 1,000 kb, and using PFGE one can separate fragment up to 5,000 kb. YAC cloning is still evolving but may soon become routine.

Once the region of DNA has been cloned, the problem remains of identifying which portion of the fragment contains the gene. One attempts to recognize the gene by identifying start and stop codons, intron/exon boundaries, and other sequences. One approach is by hybridization of genomic DNA to complementary DNAs (cDNAs) of candidate genes. These cDNAs are derived from tissue-specific messenger RNA (mRNA), which represents the tissue-expressed form of the gene. After having isolated and identified the gene, one can determine the specific mutation associated with the disease and from that devise methods to make a specific molecular diagnosis. One can also predict mutations at the level of the protein, leading to further studies on the pathophysiological mechanisms of the disease.

Detection of Mutations

As alluded to in the section of Chapter 3 describing PCR, different types of mutations can be identified by molecular scanning techniques. Deletion and insertion mutations can be identified by a special modification of the PCR (13). Mutations of much smaller magnitude, including single nucleotide mutations, may also be identified using PCR in conjunction with other techniques. The region containing the mutation is amplified by PCR using a sample of the patient's DNA and is then manipulated to identify the precise mutation. The techniques used to identify the mutation are denaturing gradient gel electrophoresis (DGGE) (14), chemical cleavage (15), and RNAse protection (16). DGGE relies on the detection of an altered mobility pattern produced by the mutation compared to the normal pattern. The chemical cleavage method uses hydroxylamine and osmium tetroxide, respectively, to detect cytosine and thymidine nucleotide mutations. These modified residues are cleaved by piperidine, and the resulting fragments can be identified by polyacrylamide gel electrophoresis and autoradiography. The RNAse protection assay is based on hybridization of RNA to the DNA of interest, followed by exposure to RNAse; digestion will only occur at sites of mismatch caused by mutations, which will show a different electrophoretic pattern compared to normal.

Application of Molecular Techniques to the Cardiomyopathies

Major advances have been made in the genetic analysis of ischemic heart disease, but such progress has not been observed in the group of diseases referred to as cardiomyopathies. These diseases are due to abnormalities directly involving the cardiac muscle either as a primary cardiac disease or secondary to abnormalities in other organs such as the thyroid or pituitary glands. Until recently there was little progress at the molecular level in the understanding of the etiologies and pathogeneses of these diseases. These patients often present with cardiac failure associated with cardiomegaly. Several of these disorders are inherited, and the associated cardiac hypertrophy is probably a secondary compensatory growth response like that which develops in response to acquired disease; however, in one of these disorders, hypertrophic cardiomyopathy, the hypertrophy appears to be primarily due to an inherited excessive growth of the ventricular septum.

In just 2 years, the application of the techniques of molecular genetics to these disorders has provided more progress in determining their etiologies than has been achieved in the previous three decades. Hypertrophic cardiomyopathy has been mapped to chromosome 14, with β-myosin heavy chain being the most likely responsible gene; the long QT syndrome has been mapped to chromosome 11; and a mitochondrial form of cardiomyopathy has been mapped to mitochondrial DNA. Duchenne muscular dystrophy, an X-linked disease caused by a defect in the protein dystrophin, is undergoing molecular analysis to elucidate the precise molecular cardiac defect. Myotonic dystrophy, which is known to affect the cardiac Purkinje system, has been mapped to chromosome 19, and the responsible gene has been isolated and identified to be a protein kinase (17). An X-linked cardiomyopathy has been mapped to the locus of the dystrophin gene and is likely to provide very pivotal information on cardiac expression of the dystrophin gene (Table 4.1).

Familial Hypertrophic Cardiomyopathy

Familial hypertrophic cardiomyopathy (HCM) is the most common cause of sudden cardiac death in young adults (18) and by the far the most common cause of sudden deaths in athletes. It is inherited in an autosomal dominant pattern with incomplete penetrance and highly variable expressivity. Pathologically, it is characterized most commonly by asymmetrical interventricular septal hypertrophy (19); however, ventricular free wall hypertrophy (20), apical hypertrophy (21), or a combination (7)

Table 4.1 Cardiomyopathies Caused By Inherited Defects

Disease	Chromosome
Hypertrophic cardiomyopathy (HCM)	14q1
Congenital long QT syndrome	11
X-linked cardiomyopathy	Xp21
Emery-Dreifuss dystrophy	Xq28
Duchenne muscular dystrophy	Xp21
Myotonic muscular dystrophy	19q13
Mitochondrial cardiomyopathy	MIT DNA

of these are also seen. Histologically, the affected myocardium has myofibrillar disarray (22) and varying degrees of fibrosis (23). The defect is one of excess cardiac growth and is manifested by myriad clinical features; thus HCM could serve as a paradigm for understanding cardiac growth as well as providing the molecular basis for many pathological features commonly present in many other diseases, such as hypertrophy, decreased compliance, altered contractility, and myofiber disarray. It is highly likely that elucidation of the primary defect will provide clues fundamental to our understanding of sarcomere assembly and cardiac growth.

In 1989, Jarcho et al. (24), using genetic linkage analysis, showed the disease to be linked to an anonymous probe on the proximal part of the long arm of chromosome 14 (14q) in a large French-Canadian family. Other studies suggested loci for HCM on chromosome 16, 18, or 2 (25–27). Subsequently, we showed genetic linkage to the same locus as that of Jarcho et al., on chromosome 14, in nine North American families (4). It is now accepted that 14q is a major locus for this disease, at least in North America. The studies purporting loci in the other chromosomes have not been substantiated. Further investigations into the families showing genetic linkage to chromosome 14 showed two mutations in the β-myosin heavy chain gene. One mutation was a unique hybrid gene formed from the fusion of the β- and α-myosin heavy chain genes with its junction at exon 27 (28). The second mutation (29) was a missense in exon 13 of the β-myosin heavy chain gene produced by a substitution of adenine for guanine, resulting in the amino acid arginine substituting for the amino acid glutamine. We have screened 60 unrelated families affected with HCM; none of them had the fused α- and β-myosin gene, and in only one family did we find the exon 13 missense mutation (30). Recently (31), at least six other mutations have been found in the β-myosin heavy chain gene, which codes for the globular protein head, and most recently we have found a unique mutation consisting of a deletion of the 3′ terminus of the gene, which encodes for the carboxy end of the β-myosin heavy chain protein (32). The myosin gene is large, consisting of more than 23 kb having 40 exons with a mRNA of greater than 6 kb. It is expected that many other mutations will be found in the myosin gene.

Studies are now underway to determine whether these mutations are expressed in cardiac tissue and ultimately to determine whether they are responsible for the cause of this disease. This may require techniques incorporating modified forms of the gene analogous to the mutant genes, such as the transgenic mouse, or homologous recombination to assess the consequences of these changes in an intact organism. There is also some concern at this time as to whether chromosome 14q1 is responsible for all forms of familial HCM. Solomon et al. (33) have claimed it is genetically heterogeneous, but no other chromosomal loci have yet been identified, and a similar claim has been made by Epstein et al. (27).

Cardiomyopathy in Duchenne and Becker Muscular Dystrophies

Duchenne and Becker muscular dystrophies (34) are caused by mutations in the gene encoding for the cytoskeletal protein dystrophin. These allelic X-linked recessive disorders are the most common and devastating of the human muscular dystrophies, affecting approximately 1 in 3,500 newborn males (35). The disease is characterized by variable but progressive degeneration and loss of function in skeletal and cardiac muscle, eventually leading to death from respiratory failure or cardiomyopathy (36,37). The cardiac manifestations also include rhythm and conduction disturbances in the heart with characteristic electrocardiographic (ECG) findings, suggesting involvement of the Purkinje system or other portions of the ventricular conduction system (38). The brain is also affected, considering that 30% of these patients are observed to have mental retardation.

The discovery of the dystrophin gene and identification of its protein product in cardiac and skeletal muscle illustrate the power of molecular biology techniques in studying diseases that influence tissue function. The dystrophin gene was first localized to the short arm of the X chromosome by linkage analysis and cytogenetically detectable defects in this region of the X chromosome in patients who had the disease (35). Subsequently, the gene was identified by hybridization of the DNA to mRNA in skeletal muscle. The RNA transcript was sequenced, from which the amino acid sequence of the protein was determined, and antibodies were produced to a previously unknown molecule now known as dystrophin (39). Studies have demonstrated that mutations in the dystrophin gene induce low-level production of a nonfunctional protein or the complete absence of dystrophin in the heart and skeletal muscle of affected patients (40, 41).

The dystrophin gene is one of the largest genes known, comprising approximately 2.5 million bp that transcribe a 14,000-bp mRNA molecule, and is expressed in striated and smooth muscle as well as brain tissue (42). In muscle tissue, the dystrophin protein has been localized to the cytoplasmic surface of the sarcolemma, and it is believed to be a cytoskeletal protein that is associated with several integral membrane glycoproteins (43). It is believed that dystrophin is involved with the regulation of intracellular calcium, since dystrophin-deficient muscle has abnormally high levels of intracellular calcium and altered calcium channel activity has been observed at the cell membrane (44). In the brain, dystrophin has been localized specifically to only the postsynaptic membrane at cortical neuronal junctions, which would suggest a more specialized function for dystrophin in the brain (42).

Studies on dystrophin mRNA in muscle and brain tissue indicate that the mRNA transcript is alternatively spliced to encode for multiple isoforms of the dystrophin protein in a developmentally and tissue-specific format, and several of these isoforms appear to be unique to the heart (45,46; R. Bies, unpublished data, 1991). This posttranscriptional modification may result in a functional diversity for the dystrophin protein. Human cardiac Purkinje fibers express dystrophin isoforms, some of which are found in the brain, that may be important for the function and integrity of these specialized cells in the heart (47).

The dystrophin mRNA present in skeletal muscle has recently been cloned as a full-length cDNA, and has been introduced and expressed in cells in tissue culture (48). This exciting advance is the first step toward developing gene therapy strategies for correcting the molecular defect in Duchenne and Becker muscular dystrophies. Our understanding of the cardiomyopathy induced by this disease is likely to be accelerated by the discovery of a mouse (called the *mdx* mouse) model for Duchenne muscular dystrophy. In this affected mouse the mutation in the dystrophin gene terminates dystrophin protein synthesis and causes some characteristic degenerative histological findings in the muscle (49,50). Similar models have been described in the dog and the chicken (51). The ability to express recombinant dystrophin in the somatic cells of the *mdx* mouse is the first hope for a possible genetic cure for this lethal disease.

Myotonic Muscular Dystrophy

Of the inherited muscular dystrophies in adults, Duchenne muscular dystrophy has the highest incidence of new cases but myotonic muscular dystrophy (MMD) is the most prevalent form. The inherited defect is autosomal dominant, with incomplete penetrance and variable expressivity. The phenotype is characterized clinically by a unique pattern of muscular weakness and wasting, myotonia, cataracts, ptosis, and frontal balding. While it is considered to be primarily a skeletal muscle disease, similar to Duchenne muscular dystrophy, it has a high incidence of serious cardiac involvement. The exact incidence of cardiac abnormalities varies from 20% to 90% of clinically affected individuals, but practically all patients with MMD have some cardiac involvement. The cardiac abnormality consists of a dilated cardiomyopathy, but pathological studies show replacement of the Purkinje system by fatty and fibrous tissue, similar to the involvement of the cardiac muscle. The abnormalities on the ECG include prolonged PR interval, intraventricular conduction defects, and bundle-branch block. Sudden cardiac death is the most common form of death in these patients.

The responsible gene has been localized to chromosome 19q13.2–13.3, flanked by the muscle creatine kinase gene and an anonymous DNA marker. It is estimated that the distance between these two markers is probably about 1 million bp, and isolation and cloning of the responsible gene are expected to be completed in the very near future (52, 53).

The Purkinje system, which appears to be specifically affected in this disease, is a highly differentiated form of muscle that conducts electrical impulses to provide uniform and rapid excitation of the heart. Identification of the protein responsible for MMD should provide insight at the molecular level into defective conduction. It will undoubtedly provide additional insight into the functions of the Purkinje system at the heart. It is unknown at this time whether the cardiomyopathy that occurs in association with MMD is secondary to the conduction abnormalities or related directly to primary abnormalities of the myocardial muscle. The recent identification of the MMD gene and its protein product (a protein kinase) has the immediate effect of making possible a precise molecular diagnosis with a more rational approach to the treatment of this disease.

Mitochondrial Cardiomyopathy

This is a disease of the mitochondria that affects the heart and skeletal muscle. The patients, however, often present with severe cardiac failure. The clinical findings include exertional dyspnea, tachycardia, general muscle weakness, and severe generalized edema with an enlarged liver and cardiomegaly. This disease is relatively newly defined, and only a few reports have been published on the genetics of this disease. It would appear that the defect is inherited maternally and is due to a mutation creating a mismatch in the anticodon stem of transfer RNA (tRNA) of the mitochondrial genome. Whether other idiopathic forms of cardiomyopathies are due to defects in the mitochondrial genome remains to be determined. While there have been few reports on mutations in mitochondrial DNA and the heart, there are other diseases known to be due to mitochondrial mutations, such as Kearns-Sayre syndrome (54). Mitochondrial DNA, as opposed to DNA in the nucleus, is transmitted exclusively by the mother and contains few, if any, noncoding sequences (introns) as well as having a slightly different genetic code. Mitochondrial

DNA is known to code for 13 proteins, all of which are components of the oxidative phosphorylation system, as well as two ribosomal RNAs and 22 tRNAs. Thus, it would be expected that most defects resulting from mitochondrial mutations are likely in some way to alter the normal oxidative metabolism and limit energy production in muscle cells with this mutation (55).

Congenital Long QT Syndrome

This syndrome is recognized for its high incidence of sudden death from ventricular tachycardia and ventricular fibrillation. Sudden death may occur in the very young or it may occur very late in life. The disease has always been of pivotal interest since it is known that drugs that prolong the QT interval are associated with an increased incidence of ventricular arrhythmias. Keating et al. (56), using linkage analysis, have mapped the locus for this gene to the short arm of chromosome 11. This is the first report of chromosomal mapping for this disorder, and they described a tight linkage to the *ras* oncogene, making it a likely candidate gene. This oncogene encodes for one of the G-proteins that is known to regulate and influence potassium channels. If, indeed, the gene encoding for *ras* oncogene is the cause for this disease, there would appear to be a rational basis for altering the electrical conduction of the heart and prolonging the QT interval. This disorder will undoubtedly shed significant light on the pathogenesis of ventricular arrhythmias as well as the cause of sudden cardiac death.

X-Linked Cardiomyopathy

X-linked cardiomyopathy (XLCM) was detailed clinically by Berko and Swift in a 63-member pedigree in 1987 (57). The males tended to develop cardiac failure in their late teens or early twenties and progressed rapidly to death. In females the disease manifests much later, in the fifth or sixth decade, and progresses very slowly, usually in a very mild form. There was no clinical evidence of skeletal myopathy except for elevated muscle creatine kinase levels. Towbin et al. (58,59) subsequently showed linkage of XLCM to the dystrophin locus at Xp21, with maximum lod scores in the most proximal (5′) portion of the dystrophin locus in this pedigree. Using an antibody to the NH_2-terminus of dystrophin, cardiac tissue showed much reduced or even absent levels of dystrophin while skeletal muscle gave normal levels. In contrast, detection with an antibody to the COOH-terminal end of dystrophin showed normal levels in heart and skeletal muscle. Furthermore, the dystrophin-associated glycoprotein described by Ervasti et al. (43) is also much reduced. XLCM thus appears to be an abnormality in the dystrophin locus causing a new phenotypic disease consisting of severe dilated cardiomyopathy without clinical evidence of skeletal myopathy. Towbin et al. have hypothesized that XLCM is due to an abnormality within the central portion of the dystrophin gene and have speculated the possible mechanism to be either 1) a point mutation in the proximal region of the dystrophin coding sequence preferentially affecting cardiac function, 2) a cardiac-specific promoter mutation that alters expression in the heart, or 3) splicing abnormalities resulting in an altered arrangement that interferes with normal cardiac expression of the dystrophin gene (59).

References

1. Ott J: A short guide to linkage analysis. In Rickwood D, Hames BD (eds): In *Human Genetic Diseases: A Practical Approach*. Washington, DC: IRL Press, 1986, pp 19–32.
2. Ott J: *Analysis of Human Genetic Linkage*. Baltimore: Johns Hopkins University Press, 1985.
3. Botstein D, White RL, Kolnick M, Davis RW: Construction of a genetic linkage map in man using restriction fragment length polymorphisms. *Am J Hum Genet* 1980;32:314–331.
4. Hejtmancik JF, Brink PA, Towbin J, et al: Localization of the gene for familial hypertrophic cardiomyopathy to chromosome 14q1 in a diverse American population. *Circulation* 1991;83:1592–1597.
5. Mares A Jr., Towbin J, Bies RD, Roberts R: Molecular biology for the cardiologist. In O'Rourke RA (ed): In *Current Problems in Cardiology* Vol. 17, St. Louis, MO: Mosby-Heart Book, Inc, 1992, pp 9–72.
6. Lander ES: Mapping complex genetic traits in humans. In Rickwood D, Hames BD (eds): *Genome Analysis: A Practical Approach*. Washington, DC: IRL Press, 1988, pp 171–189.
7. Conference proceedings, human gene mapping. *Cytogenet Cell Genet* 1988;9:46.
8. Wells RA: DNA fingerprinting. In Rickwood D, Hames BD (eds): *Genome Analysis: A Practical Approach*. Washington, DC: IRL Press, 1988, pp 153–170.
9. Coulson A, Sulston J: Genome mapping by restriction fingerprinting. In Rickwood D, Hames BD (eds): *Genome Analysis: A Practical Approach*. Washington, DC: IRL Press, 1988, pp 19–39.
10. Smith CL, Klco SR, Cantor CR: Pulsed-field gel electrophoresis and the technology of large DNA molecules. In Rickwood D, Hames BD (eds): In *Genome Analysis: A Practical Approach*. Washington, DC: IRL Press, 1988, pp 41–72.
11. Schwartz DC, Cantor CR: Separation of yeast chromosome-sized DNAs by pulsed field gradient gel electrophoresis. *Nucleic Acids Res* 1984;34:67.
12. Barlow DP, Lehrach H: Genetics by gel electrophoresis: The impact of pulse field gel electrophoresis on mammalian genetics. *Trends Genet* 1987;3:167.
13. Chamberlain JS, Gibbs RA, Ranier JE, et al: Deletion screening of the Duchenne muscular dystrophy locus via multiplex DNA amplification. *Nucleic Acids Res* 1988;16:11141–11156.

14. Myers RM, Lumelsky N, Lerman LS, et al: Detection of single base substitutions in total genomic DNA. *Nature* 1985;313:495–498.
15. Cotton RGH, Rodrigues NR, Campbell RD: Reactivity of cytosine and thymine in single base pair mismatches with hydroxlamine and osmium tetroxide and its application to the study of mutations. *Proc Natl Acad Sci USA* 1988;85:4397–4401.
16. Ganguly A, Rooney JE, Flosomi S, et al: Detection and location of single-base mutations in large DNA fragments via immunomicroscopy. *Genomics* 1989;4:530–538.
17. Fu Y-H, Pizzuti A, Fenwick R, King J, An unstable triplet repeat in a gene related to myotonic muscular dystrophy. *Science* 1992;255:1256–1258.
18. Maron BJ, Roberts WC, Edwards JE, et al: Sudden death in patients with hypertrophic cardiomyopathy: Characterization of 26 patients without functional limitation. *Am J Cardiol* 1978;41:803.
19. Epstein SE, Henry WL, Clark CE, et al: Asymmetric septal hypertrophy. *Ann Intern Med* 1974;81:650.
20. Maron BJ, Gottdiener JC, Epstein SE: Patterns and significance of distribution of left ventricular hypertrophy in hypertrophic cardiomyopathy: A wide-angle two-dimensional echocardiographic study of 125 patients. *Am J Cardiol* 1981;48:418.
21. Kereiakes DJ, Anderson DJ, Crouse L: Apical hypertrophic cardiomyopathy. *Am Heart J* 1983;105:855.
22. Davies MJ: The current status of myocardial disarray in hypertrophic cardiomyopathy (editorial). *Br Heart J* 1984;51:361.
23. Tanaka M, Fujiwara H, Onodera T, et al: Quantitative analysis of myocardial fibrosis in normals, hypertensive hearts, and hypertrophic cardiomyopathy. *Br Heart J* 1986;55:575.
24. Jarcho JA, McKenna W, Pare JAP, et al: Mapping a gene for familial hypertrophic cardiomyopathy to chromosome 14q1. *N Engl J Med* 1989;321:1372–1378.
25. Ambrosini M, Ferraro M, Reale A: Cytogenetic study in familial hypertrophic cardiomyopathy: Identification of a new fragile site on human chromosome 16. *Circulation* 1989;80(suppl II):II-458.
26. Nishi H, Himura A, Sasaki M, et al: Localization of the gene for hypertrophic cardiomyopathy on chromosome 18q. *Circulation* 1989;80(suppl II):II-457.
27. Epstein N, Fananapazir L, Lin H, et al: Genetic heterogeneity in hypertrophic cardiomyopathy: Evidence that HCM maps to chromosome 2p. *Circulation* 1990;82(suppl III):III-399.
28. Tanigawa G, Jarcho JA, Kass S, et al: A molecular basis for familial hypertrophic cardiomyopathy: an α/β cardiac myosin heavy chain hybrid gene. *Cell* 1990;62:991–998.
29. Geisterfer-Lowrance AAT, Kass A, Tanigawa G, et al: A molecular basis for familial hypertrophic cardiomyopathy: A β cardiac myosin heavy chain gene missense mutation. *Cell* 1990;62:999–1006.
30. Perryman MB, Mares A Jr, Hejtmancik JF, et al: The β myosin heavy chain missence mutation in exon 13, a putative defect for HCM is present in only one of 39 families. *Circulation* 1991;84(4):II-418.
31. Watkins H, Hwang D-S, Rosenzweig A, et al: Analyses of cardiac myosin heavy

chain gene mutations that cause familial hypertrophic cardiomyopathy. *Circulation* 1991;84(suppl II):II-418.
32. Marian AJ, Yu Q-T, Mares A, et al: Detection of a new mutation in the β-myosin heavy chain gene in a family with hypertrophic cardiomyopathy. *J Am Coll Cardiol* 1992;3(A):272A.
33. Solomon SD, Jarcho JA, McKenna W, et al: Familial hypertrophic cardiomyopathy is a genetically heterogeneous disease. *J Clin Invest* 1990;86:993–999.
34. Hoffman EP, Brown RH, Kunkel LM: Dystrophin: The protein product of the Duchenne muscular dystrophy locus. *Cell* 1987;51:919–928.
35. Emery AEH: *Duchenne Muscular Dystrophy*. Oxford Monographs on Medical Genetics, No. 15. Oxford, England: Oxford University Press, 1987.
36. Chamberlain JS, Caskey CT: Duchenne muscular dystrophy. In Appel SH (ed): *Current Neurology*, vol 10. Chicago: Year Book Medical Publishers, 1990, pp 65–103.
37. Perloff JK: Neurological disorders and heart disease. In Braunwald E (ed): *Heart Disease*, 3rd ed. Philadelphia: WB Saunders Co, 1988, pp 1782–1786.
38. Monaco AP, Neve RL, Colletti-Feener CA, et al: Isolation of candidate cDNAs for portions of the Duchenne muscular dystrophy. *Nature* 1985;323:646–650.
39. Koenig M, Monaco AP, Kunkel LM: The complete sequence of dystrophin predicts a rod shaped cytoskeletal protein. *Cell* 1988;53:219–288.
40. Bonilla E, Samitt CE, Miranda AF, et al: Duchenne muscular dystrophy: Deficiency of dystrophin at the muscle cell surface. *Cell* 1988;54:447–452.
41. Arahata K, Ishiura S, Ishiguro T, et al: Immunostaining of skeletal and cardiac muscle surface membrane with antibody against Duchenne muscular dystrophy peptide. *Nature* 1988;333:861–863.
42. Lidov HGW, Byers TJ, Walkins SC, Kunkel LM: Localization of dystrophin to postsynaptic regions of central nervous system cortical neurons. *Nature* 1990;358:725–728.
43. Ervasti JM, Ohlendieck K, Kahl S, et al: Deficiency of a glycoprotein component of the dystrophin complex in dystrophic muscle. *Nature* 1990;345:315–319.
44. Fong P, Turner PR, Denatclaw WF, Steinhardt RA: Increased activity of calcium leak channels in myotubes of Duchenne human and mdx mouse origin. *Science* 1990;250:673–676.
45. Nudel U, Zuk D, Einat P, et al: Duchenne muscular dystrophy gene product is not identical in muscle and brain. *Nature* 1989;337–76–78.
46. Feener CA, Koenig M, Kunkel LM: Alternative splicing of human dystrophin mRNA generates isoforms at the carboxy terminus. *Nature* 1989;338:509–511.
47. Bies RD, Friedman D, Roberts R, et al: Expression and localization of dystrophin in human cardiac Purkinje fibers. *Circulation* 1992;86.
48. Lee CC, Pearlman JA, Chamberlain JS, Caskey CT: Expression of recombinant dystrophin and its localization of the cell membrane. *Nature* 1991;349–334–336.
49. Bulfield G, Siller WG, Wight PAL, Moore KJ: X chromosome-linked muscular dystrophy (*mdx*) in the mouse. *Proc Natl Acad Sci USA* 1984;81:1189–1192.
50. Capecchi MR: Altering the genome by homologous recombination. *Science* 1989;244:1288–1292.

51. LeMaire CR, Heilig R, Mandel JL: The chicken dystrophin cDNA: Striking conversation of the C-terminal coding and 3′ untranslated regions between man and chicken. *EMBO J* 1988;7:4157–4162.
52. Epstein HF, Ashizawa T, Perryman MB, et al: Multiple molecular genetic strategies to identification of the myotonic dystrophy locus. *J Neurosci* 1990;98:39.
53. Perryman MB, Bachinski L, Cortez MD, et al: Molecular genetics of myotonic dystrophy. *J Cell Biochem* 1991;15C:155.
54. Moraes CT, DiMauro S, Zeviani M, et al: Mitochondrial DNA deletions in progressive external ophthalmoplegia and Kearns-Sayre syndrome. *N Engl J Med* 1989;320:1293–1299.
55. Zeviani M, Gellera C, Antozzi C, et al: Maternally inherited myopathy and cardiomyopathy: Association with mutation in mitochondrial DNA tRNA. *Lancet* 1991;138:143–147.
56. Keating M, Atkinson D, Dunn C, et al: Linkage of a cardiac arrhythmia, the long QT syndrome, and the Harvey *ras*-1 gene. *Science* 1991;252:704–706.
57. Berko BA, Swift M: X-linked dilated cardiomyopathy. *N Engl J Med* 1987;316:1186–1191.
58. Towbin JA, Brink P, Gelb B, et al: X-linked cardiomyopathy: Molecular genetic evidence of link to the Duchenne muscular dystrophy locus. *Pediatr Res* 1990;27:25A.
59. Towbin JA, Zhu XM, Gelb BD, et al: X-linked dilated cardiomyopathy (XLCM): Molecular pathogenesis. *Pediatr Res* 1991;29:25A.

Chapter **5**

Molecular Biology of Contractile and Cytoskeletal Proteins

The Contractile Apparatus

Historically, myocardium held little interest for molecular biologists because it appeared to be genetically static in the developed organism and capable only of stereotypical responses to diverse stimuli. However, the seminal observation that myosin could undergo isoform transitions in the overloaded heart led to molecular investigation of the myocardium, which revealed that most sarcomeric proteins are products of multigene families with a complexly regulated pattern of expression. Thus, an understanding of the structure of the cardiac sarcomere, and the physiological and pathological isoform transitions that occur therein, serves as a useful paradigm for molecular biology of the heart. This chapter provides an outline of sarcomeric structure and summarizes new knowledge concerning the regulation of the genes encoding the contractile apparatus (see refs. 1–12 for reviews).

Structure of the Cardiac Sarcomere

Cardiac ventricular muscle cells contain a centrally located nucleus and form a functional syncytium through end-to-end connections called *intercalated discs*. Unlike skeletal muscle, they do not form multinucleated myotubes. The myocyte is made up of a large number of cross-banded strands called *myofibrils* that are arranged in register within a given cell, giving cardiac muscle its typical striated appearance. Myofibrils themselves are made up of longitudinally aligned sarcomeres that are fundamentally similar in structure in both cardiac and skeletal muscle. Under light microscopy, myofibrils consist of alternating dark and light bands that are termed *A-bands* (anisotropic) and *I-bands* (isotropic), respectively (Fig. 5.1). The A-band contains a centrally located lighter zone called the *H-band* that is bisected by a darker *M-line*. The I-band is similarly bisected by a dark *Z-line*. The sarcomere consists of the portion of the myofilament between two adjacent Z-lines and in the resting state

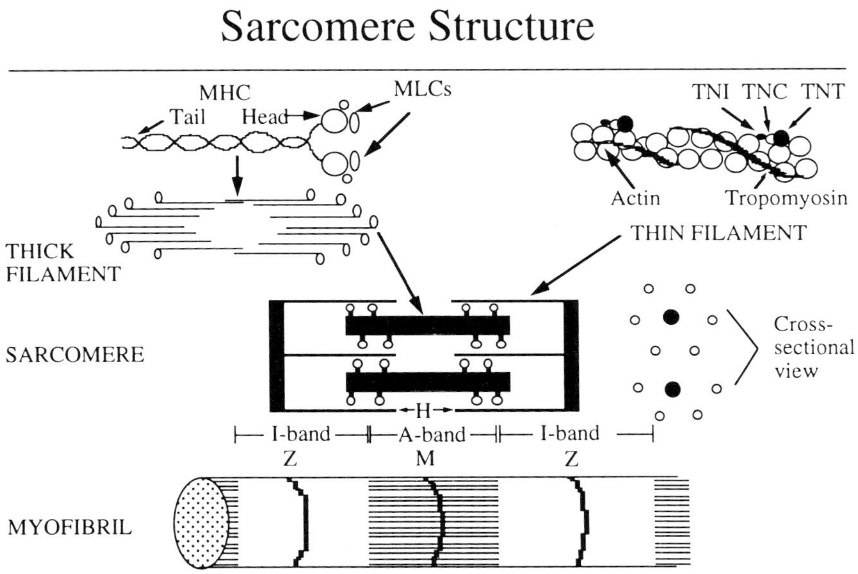

Figure 5.1 Sarcomere structure: *MHC*, myosin heavy chain; *MLC*, myosin light chain; *TNI*, troponin I; *TNC*, troponin C; *TNT*, troponin T; *H*, H-band; *M*, M-line; *Z*, Z-line.

is approximately 2.2 μm in length. During myocardial contraction, the A-band does not change in size whereas the total sarcomeric length can decrease to approximately 1.6 μm.

Electron microscopy shows the sarcomere to be an interdigitation of thick and thin filaments. The thick filaments are approximately 1.5 μm long and constitute the A-band. The thin filaments are approximately 1 μm long and originate from the Z-line, terminating within the A-band at the boundary of the H-band. Thus, the darkness of any given sarcomeric band represents overlap, or lack thereof, of the two types of filaments. The stoichiometry of thin to thick filaments is approximately 6:1; a cross section through the A-band outside of the H-band reveals a central thick filament surrounded by six hexagonally arranged thin filaments.

Constituents of the Thick Filament

The thick filament consists essentially of the protein myosin. Myosin is a fibrous protein that has the capacity both to hydrolyze adenosine triphosphate (ATP) upon activation by actin in the presence of magnesium and to bind directly with actin itself. Each molecular of myosin consists of two myosin heavy chains (MHC) and two pairs of myosin light chains (MLCs). Each MHC consists of a long α-helical segment and a globular NH_2-terminus (Fig. 5.1). In the intact molecule, the two helical portions of the MHC coil around each other, forming a single, rod-like tail with two heads. MLCs are associated with each head, which contains the enzymatic activity and the site of actin binding. Within a given thick filament there are approximately 400 myosin molecules, which are arranged in an anti-parallel fashion, resulting in a bipolar arrangement with myosin heads at each end. The M-line of the sarcomere is the region where the myosin rods come together and are linked by a variety of proteins, including creatine phosphokinase, M-protein, and C- and H-proteins, and the resulting M-line is surrounded by a sleeve of the protein myomesin.

Constituents of the Thin Filament

The thin filament consists of two chains of globular actin molecules arranged in a double helix. This helix contains approximately 13.5 actin molecules per turn and is associated every 38.5 nm with a troponin-tropomyosin complex. The troponin-tropomyosin complex is inserted into the two grooves formed by the actin helix (Fig. 5.1). Troponin consists of three separate protein components: troponin C, which binds calcium; troponin I, which binds actin; and troponin T, which stabilizes the entire complex and binds to tropomyosin. Tropomyosin itself is a rodlike protein made up of two dimeric chains arranged in a superhelix. It should be noted that the thin filaments are also polar, with polarity determined by their interaction with the myosin heads.

In addition to the essential proteins constituting the thin filaments, a variety of other proteins exist within the sarcomere, including those comprising the *Z-disc*, which provide mechanical stability and continuity to the myofibril, and a variety of cytoskeletal and filamentous proteins that are described in the second portion of this chapter.

Cardiac Contraction

A full description of the complex process of excitation-contraction coupling exceeds the scope of this text. However, the critical contribution of steric changes in the contractile proteins to cardiac contraction is described briefly (Fig. 5.2).

Cardiac Muscle Shortening with Actin-Myosin Interaction

In the resting, noncontractile state, cardiac contraction is prevented by the presence of tropomyosin over the site of the actin strand required for interaction with myosin. A cardiac action potential results in depolarization of the sarcolemmal membrane, with a resultant rise in intracellular calcium (10^{-7} to 10^{-5} mol/liter), in large part from stores within the sarcoplasmic reticulum. Three moles of calcium bind to each troponin C molecule, resulting in the inhibition of the binding of troponin I to actin. In addition, steric conformational changes occur in tropomyosin, enabling actin to contact the myosin head. In the presence of ATP and magnesium, the myosin head "swivels" such that the thin filaments are pulled toward the center of the sarcomere. Following completion of

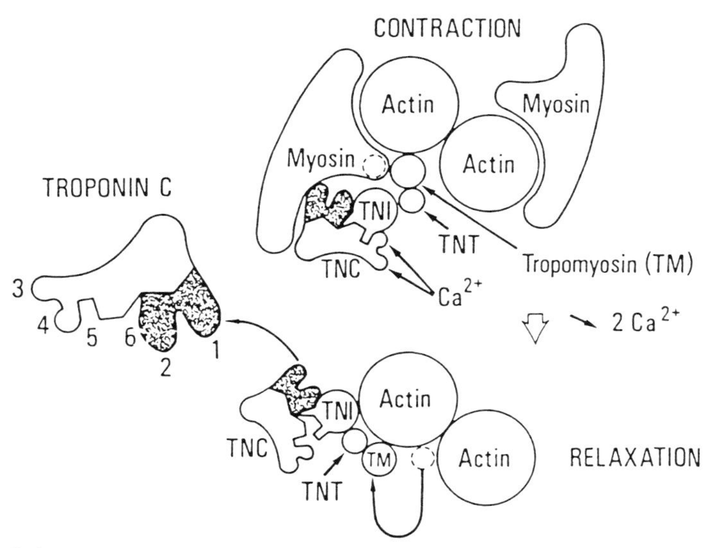

Figure 5.2 Scheme of the steric model of contraction in a skeletal sarcomere. Transverse section of a sarcomere showing the movement of the tropomyosin molecule (*TM*) during relaxation. TM leaves the actin-actin groove to prevent actin-myosin interaction. Fixation of calcium on the specific calcium binding sites 3 and 4 of TNC enhances the affinity of TNC for TNI (site 5). The two nonspecific sites of TNC, 1 and 2, compete with Mg^{2+}; this is why they are nonspecific.

this "rowing motion," the myosin head ejects the ATP hydrolysis products, adenosine diphosphate (ADP) and inorganic phosphate, and then binds another molecule of ATP, which results in disassociation. As long as intracellular calcium levels remain high enough, the myosin head is then free to interact with another site on the actin molecule, resulting in further shortening of the sarcomere. Thus the shortening of cardiac muscle results from the relative sliding of the thin actin filament along the thick myosin molecule. At the end of diastole when intracellular calcium levels fall, the troponin-tropomyosin complex returns to its initial position on the actin molecule, which prevents further actin-myosin interaction, with resultant relaxation. Thus, the cardiac contractile apparatus is controlled by a "repressor" mechanism that is released by elevation of intracellular calcium.

The force generated within cardiac muscle is directly related to the quantity of calcium that is bound to troponin C. In addition, the specific isoforms of the MHC that are present within cardiac muscle determine the ATPase activity (see later). Finally, various contractile proteins can be phosphorylated, presumably regulating their function in the contractile process. Not only troponin T and troponin I are phosphorylatable the MLCs can be similarly modified, which may produce changes in the enzymatic activity of myosin itself. Structural changes within the contractile proteins, therefore, account for not only cardiac contraction itself but likely, in part, regulation of the state of cardiac contractility.

Thick Filament Isoforms

The demonstration that structurally altered myosin accounted for changes in ATPase activity and cardiac contractility in a variety of pathophysiological states suggested the existence of a variety of cardiac isomyosins. Indeed, these isomyosins exist as a result of variability in the composition of their MHCs, their MLCs, or both. Although this was originally determined primarily by immunological methods, molecular genetics has provided a clearer understanding of the diversity of myosin proteins and their regulation.

Isoforms of MHC in Cardiac Muscle

In a given species, at least seven genes exist that encode for MHC proteins. These genes are expressed in mammalian striated muscle and are regulated through development and by a variety of stimuli. Two isoforms are expressed in mammalian cardiac muscle, an α-MHC and a β-MHC; the latter is also expressed in slow adult skeletal muscle. These proteins are produced by two separate genes that have been cloned in humans, rats, rabbits, and the Syrian hamster. The human gene is present on chromosome 14. In both rats and humans, the two genes are on a contiguous segment of DNA with the β-MHC gene several kilobase pairs upstream from the α-MHC gene. There is about a 20% to 30% homology between the two genes, with a homology of 90% to 95% in the coding sequences. Indeed, the genes appear to diverge primarily in the COOH-terminal peptide and the 3′ noncoding region. Intriguingly, structural mutations of the MHC genes in man have been associated with familial hypertrophic cardiomyopathy in two kindred. These mutations include mutations within the coding region of the β-MHC and an α/β-MHC fusion gene.

Within the cardiac ventricle three ventricular isomyosins exist as a result of combinatorial permutations of the α- and β-MHC. Thus, V1 myosin consists of two α-MHCs, V2 is a heterodimer of α and β, and V3 consists of two β-MHCs. These three isomyosins hydrolyze ATP at different rates as a result of their different MHC compositions. Thus V1 myosin has approximately a four- to fivefold higher ATPase activity than V3 myosin. Within cardiac muscle, for any given activation degree, the rate of shortening of muscle is directly dependent upon the myosin ATPase activity. Therefore V1 myosin results in a rapid rate of shortening, which confers a lower contractile efficiency than that associated with the V3 isoform.

MLC Isoforms Expressed in Myocardium

Three MLC isoforms exist, MLC 1 and 3, which dissociate at high pH (alkali MLC), and a phosphorylatable MLC 2 (regulatory MLC). Cardiac muscle expresses only MLC 1 and 3, whereas MLC 2 is expressed exclusively in skeletal muscle. Three MLC 1 genes exist, two of which are expressed in cardiac muscle, one in the adult ventricle (MLC 1v) and one in the adult atria (MLC 1a). Similarly, there appear to be two distinct genes for MLC 2, also an atrial and a ventricular form. Interestingly, the ventricular form of MLC 2 is nevertheless expressed in the interatrial septum.

Regulation of Thick Filament Protein Genes in Normal Development

Throughout embryonic life, all species of mammals studied to date express predominantly β-MHC in the ventricles and α-MHC in the atria (Table 5.1). Near the end of gestation, there is the appearance of α-MHC in the ventricles and β-MHC in the atria. Thereafter, the relative expression of these two forms in the cardiac chambers exhibit species-specific differences. In small rodents such as rats and mice, at the age of 3 weeks the ventricles express almost exclusively α-MHC, with minimal β-MHC expression. Thereafter there is the slow accumulation of β-MHC expression such that up to 50% of the ventricular and 15% of the atrial expression is β-MHC at senescence. In contrast, the neonatal rabbit ventricle transiently expresses α-MHC, but by maturity expression is almost exclusively β-MHC. Finally, in larger animals, including humans, there is little increase in the amount of α-MHC expression in the ventricle at birth and the predominant ventricular form is β-MHC throughout life. In the atria, despite some regional expression of β-MHC, α-MHC appears to predominate.

In early embryonic life the ventricular and atrial forms of MLC 1 are expressed in the appropriate cardiac chambers. However, there is moderate expression of atrial MLC 1 in the embryonic ventricle that diminishes throughout development. Also, at least in bovine myocardium, the ventricular form of MLC 2 is expressed in the embryonic atria.

Table 5.1 Myocardial Expression of Myosin Heavy Chain Isoforms

	Fetus	Adult	Pressure-Overload	Hyperthyroid	Hypothyroid
Rat ventricle	β	α	β	α	β
Human ventricle	β	βα	β	?	?
Human atrium	βα	βα	βα	?	?

β = β-myosin heavy chain; α = α-myosin heavy chain
(Adapted from Swynghedauw B: *Cardiac Hypertrophy and Failure*.)

Pathophysiological Regulation of Thick Filament Isoforms

The thick filament protein composition is markedly altered in a variety of pathophysiological states. In rodent models, the induction of pressure overload cardiac hypertrophy is associated with reexpression of a fetal program of gene expression including marked induction of β-MHC in the adult ventricle. This induction has been demonstrated both at the level of protein expression and by in situ hybridization at the level of messenger RNA (mRNA). In fact, there is some regional variability to this expression, with a higher degree of induction in the subendocardium, which is exposed to higher wall stress, and around cardiac vasculature. There is a concomitant, although somewhat temporally dissociated, suppression of α-MHC expression. A similar transition occurs in the overloaded human atria, and the degree of reexpression of β-MHC in this situation appears to correlate well with load as measured by echocardiographic changes in left atrial size.

In the ventricular tissue of larger animals, including humans, where β-myosin is already the predominant form, transitions do occur but they are less marked. There is loss of any remaining α-MHC expression with some increase in expression of the β-MHC form. However, as outlined in Chapter 6, a fetal program is induced in human ventricular cardiac disease, including reexpression of the normally fetal atrial MLC 1.

Other pathological situations are associated with differential induction of thick filament isoforms. The cardiac enlargement associated with physical training leads to an accumulation of α-MHC. Similarly, hyperthyroidism results in predominant expression of α-MHC in rodents, with loss of any residual β-MHC expression, in contrast to the opposite effect seen with hypothyroidism. Although thyroid hormone is, in fact, the canonical regulator of MHC expression, many of the transitions seen with pressure overload in vivo appear to be independent of such hormonal control (see Chapter 6).

Although, experimentally, isomyosin switches are associated with changes in ATPase activity and maximum rate of shortening, the functional consequences of these transitions in intact animals are less clear. In particular, the minimal transitions that occur in humans, and those transitions that occur in the atria, may be of only limited adaptive value. It has been suggested that these transitions represent a program of shared regulation rather than a true adaptive response to physiological changes.

Thin Filament Isoforms

As with the thick filament, the proteins that comprise the thin filament are products of multigene families. The greatest fund of knowledge is available for expression of the α-actins in myocardium.

α-Actins Expressed in Myocardium

Two sarcomeric actins are expressed in myocardium and are encoded for by two separate genes, skeletal α-actin and cardiac α-actin. The proteins encoded by these genes are structurally very similar, differing by only four residues out of 375 amino acids. Thus biochemical characterization and separation have been difficult. The two genes are also structurally very similar, differing primarily in their 3′ noncoding region, which enables detection by specific hybridizing complementary DNA probes. Each gene contains five introns of varying sizes, and there is a high degree of homology between species. A third α-actin, smooth muscle α-actin, is also expressed in embryonic myocardium. This gene is normally expressed at high levels only in nonstriated smooth muscle.

Tropomyosin Isoform Diversity

Tropomyosin molecules normally exist as a supercoiled dimer, with α and β subunits. These subunits are encoded by two separate genes that are capable of undergoing alternative splicing to produce a variety of peptides. The α-tropomyosin gene can produce three distinct mRNAs, α_1, α_2, and α_3. Only α_2 is present in ventricular tissue, whereas α_1 and α_2 are present in skeletal muscle and α_3 is present in smooth muscle. The β-tropomyosin gene can produce mRNAs of 2.5 kilobase or 3 pairs (kb) in size. The smaller protein is also produced in nonmuscular systems, and the larger is found in both skeletal muscle and the heart. Thus the tropomyosins have a pattern of expression and organization that is distinct from that of either the myosins or the actins. The functional consequences of the different proteins produced by such alternative splicing are unclear.

Troponin Isoforms in the Heart

Less information is available about the isoforms of the three troponins expressed in cardiac muscle. It appears that a minimum of two isoforms exist for each of troponin T, troponin I, and troponin C. At least ten isoforms of troponin T are constructed by alternative splicing in skeletal muscle. Many of the potential RNAs have been characterized in the rat, but alternative splicing of this gene has also been shown in humans. Immunological analysis has demonstrated two isoforms in the rat heart; a larger form is present in the fetal ventricle that undergoes a transition to a smaller form in adulthood. Similarly, cardiac troponin I is a larger molecule than that found in skeletal muscle. Both isoforms are phosphorylatable but apparently at different residues in the cardiac versus skeletal protein. It should be noted that phosphorylation of troponin I appears to decrease the sensitivity of cardiac fibers to calcium and has been postulated to account for the improved relaxation seen with β-adrenergic agonists. Only this single form of troponin I has been described in cardiac muscle. Analogous to β-MHC, the troponin C isoforms identified in the heart and slow skeletal muscle of the rabbit are identical. Although only a single isoform exists in the heart, another alternatively spliced isoform, with different calcium binding properties, exists in fast skeletal muscle.

Regulation of Thin Filament Isoforms During Development

As seen with the MHCs, there is some heterogeneity of expression of isoactins between rodents and larger mammals. In avian and rodent hearts, it appears that the smooth muscle isoform is expressed in the early embryo and perhaps heralds the onset of cardiac differentiation. The expression of this gene is rapidly lost and supplanted by both skeletal and cardiac α-actin. The skeletal form predominates in the fetal heart, but a transition in the newborn period results in a predominance of cardiac α-actin. In contrast, there is persistent and variable expression of skeletal α-actin in the ventricles of humans. It appears, however, that cardiac α-actin RNA is the predominant form in the atrium.

In rodents the α-isoform of tropomyosin is the only form present in the cardiac ventricles. In contrast, in larger mammals, including humans, there is approximately 20% expression of the β-isoform. Unlike other sarcomeric proteins, there appears to be little difference between atria and ventricles. As alluded to above, troponin T undergoes transitions between fetal and adult rodent ventricles. There is also some suggestion of a transition, at least in the canine myocardium, in troponin I expression. As of yet no alterations in troponin C expression through development have been seen.

Regulation of Actin and Tropomyosin Gene Expression by Pressure Overload

Following aortic coarctation in the rat, there is the rapid and transient reexpression of skeletal α-actin. Unlike the reexpression of β-MHC, the skeletal α-actin reinduction is a more diffuse process and follows a different time course. Recent data also demonstrate the reexpression of the nonsarcomeric smooth muscle α-actin in overloaded rat ventricular myocardium. In diseased human ventricles, there is marked heterogeneity of expression of the sarcomeric α-actins and no predictable reexpression of the skeletal α-actin gene. The consequences of reexpression of skeletal α-actin in overloaded myocardium are unclear. It should be noted, however, that two of the amino acid differences between these two proteins are at the site of interaction with myosin heads.

The only other transition is the thin filament documented in overloaded myocardium to date involve the reexpression of β-tropomyosin in the rat ventricle. Transitions in the troponins under pathological conditions have yet to be studied.

The Cytoskeleton

The cytoskeletal elements of the cell form the basic structural elements that maintain cellular integrity and shape. The ability of the cell to withstand shear forces during mechanical loading is determined by the type of cytoskeletal proteins in a cell and how they interact. In the myocardium, pressure load, volume load, hypercontractile states, and cardiac failure impose a variety of forces that are exerted on cytoskeletal and contractile elements of the heart, which may result in the pathophysiological effect.

The cytoskeletal is made up of a number of different protein filaments that extend throughout the cytoplasm in a complex network that helps organize other molecules and organelles within the cell (13). The cytoskeletal is not a passive structure because it must adapt to changes in cell shape, movement, and division, as well as carry out a transport function for organelles and vesicles within the cell. These diverse functions of the cytoskeletal depend on four principal types of cytoskeleton elements: microfilaments, microtubules, intermediate filaments, and membrane-associated filaments (linking the first three to the membrane surface).

In muscle tissue additional *accessory proteins* are also associated with maintaining the architecture of the myofibril by linking the sarcomeric contractile elements to themselves and then to the cell membrane so that their force of contraction can be imparted to the adjacent cell (14). Sarcomeric elements associated with the myosin thick filament include the protein called *titin*, which is approximately 2,500 kD in size, that forms an elastic network linking the thick filaments to the Z-disks. In the central portion of the sarcomere *myomesin*, a 185-kD protein present at the central M-line of the muscle thick filament, and C-protein, a 140-kD molecule found in distinct stripes on either side of the thick filament M-line, are involved in maintaining the sarcomeric architecture. Actin filaments within the sarcomere are linked to the Z-disk via an elongated inextensible protein called *nebulin*, a 600-kD protein oriented parallel to actin filaments that is attached to the Z-disk. α-Actin, a 190-kd protein, acts within the sarcomere to bundle actin molecules and link these filaments together in the region of the Z-disk. As is discussed later, α-actinin is also associated with other cytoskeletal structures such as the actin filaments that are associated with the cell membrane.

Actin Filaments

Actin filaments are polymerized forms of actin monomers that form a cross-linked network within the cell cytoplasm that becomes more dense just under the plasma membrane, constituting the *cell cortex* (15). Actin filaments may also be involved in cell movement or changes in cell shape by asymmetrical growth, which involves further polymerization at one end of the actin filament, called the plus end. Depolymerization or shortening of the actin filament may also occur (16). The rigidity of the actin network is also under dynamic control by the degree of cross-linking. The most abundant cross-linking protein is called *filamen*. Filamen is a flexible molecule composed of two polypeptide chains joined head to head with a binding site for actin (17). Other molecules make it possible for the actin network to adopt a more fluid state. One of these proteins is called *gelsolin*, a 90,000-kD protein that, in the presence of calcium, severs actin filaments and forms a cap on the end of the actin, thus breaking up cross-linked networks of actin filaments (18). Approximately 50% of the actin molecules in most cells are unpolymerized, existing either as free monomers or in complexes with other proteins. The dynamic equilibrium between unpolymerized actin molecules and actin filaments allows for an adaptive mechanism of the cytoskeleton to different cell conditions and stresses. As is discussed later, the cell cortex is a molecular complex at the cytoplasmic surface of the cell membrane where actin filaments are in association with specific proteins, which in turn are associated with other proteins embedded into the cell membrane. The *spectrin* and *ankyrin* complex is one such cytoskeletal network that provides mechanical support for the cell membrane and allows for attachment of the cytoplasmic actin filaments (19). Spectrin-related proteins such as α-actinin and dystrophin also form membrane cytoskeletal networks similar to that of spectrin.

Microtubules

Microtubules are long, filamentous structures found both in the cell nucleus and in the cytoplasm. Microtubules are involved in a vast array of cellular functions, including localization of the endoplasmic reticulum and Golgi apparatus in cells (20,21), vesicular movement, cytoplasmic transport, and centromere formation, and chromosomal separation during cell division, as well as in specialized structures associated with the cell, such as nerve axons, cilia, and flagella-type structures in mobile cells (22). Microtubules, similarly to actin filaments, display a high degree of plasticity in their ability to assemble and disassemble rapidly within the cell. Microtubules are formed by the association of two proteins subunits, α-tubulin and β-tubulin, which are each encoded by several genes (23). The tubulin proteins are highly conserved across species, suggesting their general importance during cellular evolution. α and β subunits of tubulin associate in an alternating pattern, creating a long strand called the *protofilament*. Thirteen protofilaments associated in parallel around a central core that appears to be empty form a complete microtubule (24). These microtubules then associate as doublets or triplets to form different structures such as cilia and centrioles, or they may remain as single microtubules arranged in highly regulated structural patterns within the cell cytoplasm. Microtubules are seen in the greatest density around the nucleus and radiate out into the cellular periphery as a fine lacy threads.

Microtubules can be depolymerized with the drug colchicine. The cell origin of microtubules is seen when colchicine is added to a cell culture (25). The microtubules can then be stained with antitubulin antibody and are found to be generated around a structure called the *microtubule organizing center*. Initially they form a starlike structure called an *aster* in the perinuclear region. They then elongate toward the periphery until the original distribution is established. In vertebrates, this microtubular organizing center is called the *centrosome*.

Microtubules have been shown to be modified posttranslationally for their function in different areas of the cell. Tubulin acetyltransferase acetylates a specific lysine of the α-tubulin subunit, which is the most common type of tubulin found in ciliary structures (26). The second modification is decarboxylation of the terminal tyrosine residue in α-tubulin, which is the dominant type of tubulin found in older microtubules that survive the normal microtubular turnover (27).

There is a group of proteins called *microtubule-associated proteins* that are believed to enable interactions between other cell components and the microtubules, as well as conferring stabilization of microtubules against disassembly (28). There are two major classes of microtubule-associated proteins: 1) high-molecular-weight proteins in the range of 250 kD, and 2) tau proteins with a molecular mass of approximately 50 kD. Many of these microtubule-associated proteins bind selectively to microtubules and provide permanent links to other cell components, including other parts of the cytoskeleton and selected organelle membranes.

Vesicle transport within cells has been shown to be controlled by movement along microtubules. Two microtubule-associated proteins that are capable of hydrolyzing ATP are responsible for this movement. *Kinesin* is an ATPase that uses the energy of ATP hydrolysis to move cytoplasmic vesicles unidirectionally along a microtubule at the rate of 0.5 to 2 mm/s. This movement occurs only in an outward direction toward the plus ends of the microtubules. A second protein called *cytoplasmic dynein* is responsible for transport in the other direction, from the terminus back toward the cell body (29). The kinesin and dynein motor proteins have been best studied within neurons of mammalian brain; however, they are not confined to neural tissues. Kinesin has been found at the membrane of the endoplasmic reticulum, which stretches outward from the centrosome along microtubules (20). There also appear to be dynein-like proteins associated with the Golgi apparatus that, in contrast, become localized near the centrosome (21).

Intermediate Filaments

Intermediate filaments are so called because their size, which is between 8 and 10 nm in diameter, is intermediate between the thin and thick filaments in muscle cells. Desmin is an intermediate filament found in muscle cells that appears to tie the edges of Z-disks together (30). Another example is vimentin filaments, which are found in fibroblasts and many other cell types near the cytoplasmic surface of membrane contact points between cells (31). Intermediate filaments are ropelike polymers that are thought to play a structural or tension-bearing role in the cell. The diverse intermediate filaments appear to be specific for each cell type in which they are found. Intermediate filaments are found throughout the cell membrane. They tend to form a basketlike structure around the cell nucleus and extend out in arrays throughout the cell cytoplasm. The primary structure is a coiled coil dimer (32) that contains four α-helical regions shared by all intermediate filaments. The two polypeptide chains of an intermediate filament homodimer line up in parallel, giving a central rodlike domain and two globular domains at each end. Globular domains vary in size and differ in each of the intermediate filament proteins. Assembly of these intermediate filaments has been shown to be partially controlled by phosphorylation by protein kinases, which causes filament disassembly into smaller units (30). When modulating the properties of the intermediate filament, the variable regions determine its ability to associate not only with itself but also with other cellular components such as the plasma membrane, microtubules, and other major elements of the cytoskeleton, such as actin filaments and proteins that make up the cell cortex, the cytoplasmic surface of the cell membrane (33).

Both skeletal and cardiac myocytes contain desmin filaments, which are the major proteins forming the desmosome connecting one muscle cell to the other. These important structures transmit force of sarcomeric contraction between adjacent cardiac and skeletal myocytes. There are no known genetic defects that involve intermediate filaments. It is likely that disruption of the ability to generate desmin molecules would confer a lethal phenotype. However, many cells can survive without cytoplasmic intermediate filaments, such as glia cells, which do not contain these molecules. In addition, fibroblasts that have had their intermediate filaments disrupted by intracellular injection of antibodies appear to have normal cell organization and behavior. These studies imply that the function of intermediate filaments is not critical in all cell types (26).

Membrane-Associated Cytoskeleton

While the cytoskeleton is known to form a network of supportive structural proteins beneath the cell membrane, it is important to recognize that cytoskeletal elements appear to modulate cellular physiology. A large number of ion channels and cell surface receptors have been shown to be bound and localized to the membrane surface via specific interaction with cytoskeletal proteins. Spectrin is one such molecule that is localized to the cytoplasmic surface of the cell membrane, and, via its interaction with ankyrin, binds to a number of important membrane complexes, including the voltage-dependent sodium channel, Na^+/K^+-ATPase, and the acetylcholine receptor (34). Dystrophin is a spectrin-like molecule that is also localized to the cytosolic membrane surface and is known to bind a transmembrane glycoprotein complex (35). The importance of dystrophin is illustrated in patients with Becker or Duchenne muscular dystrophy, in whom mutations in the dystrophin gene cause cardiomyopathy and conduction disease in addition to the skeletal muscle weakness (36,37). The observation that the cytoskeletal proteins are not generic, but rather produce cardiac-specific isoforms, is intriguing in terms of evolutionary adaptation to unique aspects of cardiac physiology (38). This section briefly outlines the organization of these membrane-associated cytoskeletal elements in the cell using information from a variety of biochemical studies. In addition, recent molecular analyses are discussed in terms of isoforms of these proteins that may tailor them for specific function in the heart.

α-Actinin and Intracellular Actin Filaments

α-Actinin and intracellular actin filaments may form bundles that bind to the plasma membrane in a way that allows them to pull on the extracellular matrix or another cell via linkage at a focal contact point at the cell membrane. These focal contacts between cells are mediated by membrane-bound receptor molecules of the *integrin* family, such as the fibronectin receptor (Fig. 5.3).

α-Actinin is a rodlike structural cytoskeletal element homologous to spectrin that forms cross-links between cytoplasmic "stress fibers" in the cytoplasmic architecture and the membranous integrin molecules at the focal contact sites (39). An example of the architecture of this group of cytoskeletal elements is the linkage of intracellular actin fibers with *fibronectin* in the extracellular matrix. Fibronectin is a major glycoprotein that is bound to the plasma membrane in many cell types, including muscle, through an integral membrane protein called the *fibronectin receptor*. These receptors are positioned at points of focal contact in a cluster that is organized by the interaction of α-actinin with actin. α-Actinin creates cross-links between a number of actin filament stress fibers at their ends for attachment in a region of focal contact at the cell membrane. The α-actinin cross-linking of actin filaments provides a binding site for two intermediate molecules, vinculin and talin, which then bind and anchor the α-actinin stress filament ends to the point of focal contact via the fibronectin receptors.

As mentioned previously, α-actinin also serves an important role in

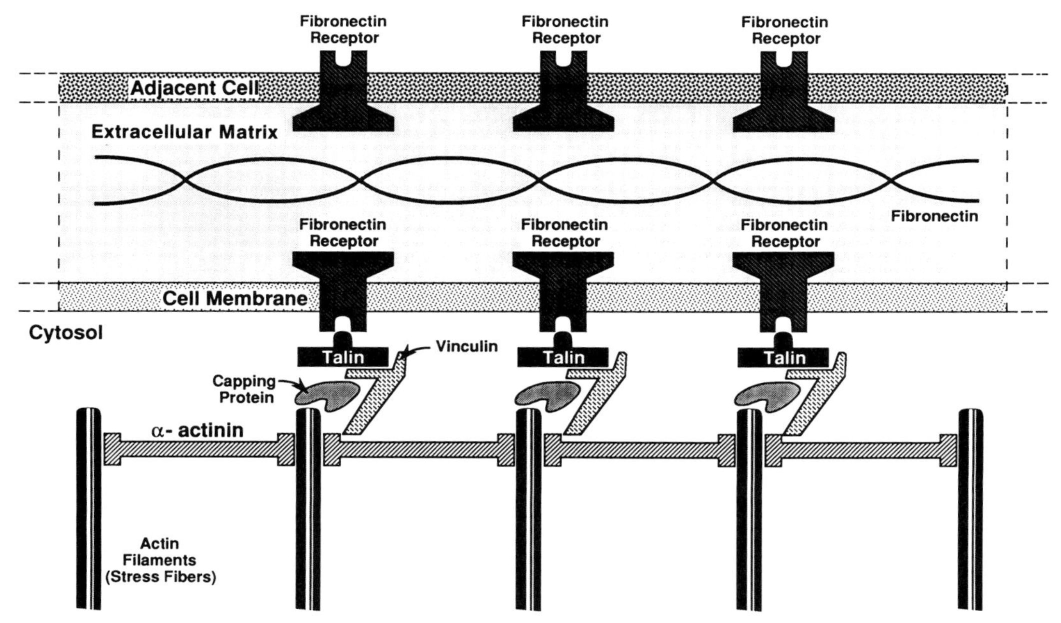

Figure 5.3 Schematic representation of proposed interactions of the cytoskeletal protein α-actin with actin stress filaments and receptor-associated proteins at the cell membrane. This example illustrates how vinculin, talin, and capping protein may associate with the fibronectin receptor, which binds to fibronectin and contributes to the transmission of intracellular adhesion forces and the cytoskeleton of the cell.

the organization of actin filaments in sarcomeric units in the muscle. α-Actinin has been localized to the Z-disk, where it is believed to link the ends of the actin filaments as a stable sarcomeric structure that can transmit the contractile force.

The cell-cell junctions in epithelial cells are also regulated by the cytoskeletal elements, which form cell-cell attachments called *adhesion belts* that help join epithelial cells into sheets. This occurs via focal contact bundles of actin filaments that interact across adjacent plasmid membranes via cell-cell junctions that contain α-actinin and vinculin but not talin, suggesting a somewhat different attachment of the actin filament to the plasmid membrane in this cell type.

Spectrin and Ankyrin

Spectrin is one of the first cytoskeletal proteins to be identified and studied. It was originally isolated from red blood cell membranes, where it was found to be an antiparellel heterodimer of α and β subunits associated noncovalently to the cytoplasmic surface's cell membrane (34). Spectrin is expressed as isoforms of a molecular mass range of 220 to 260 kD. It is found in a number of different tissues, including brain, skeletal muscle, and heart, and has been extensively studied in terms of its

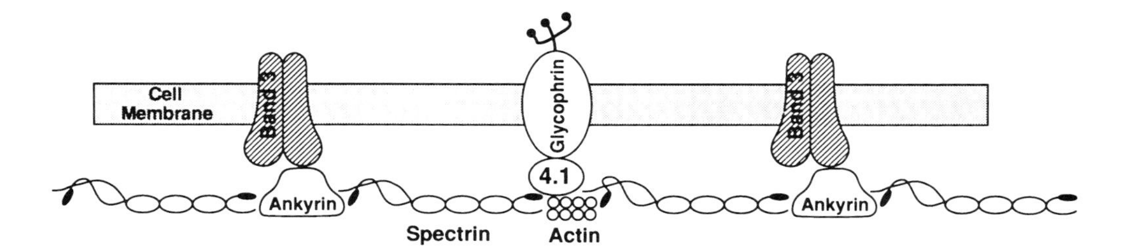

Figure 5.4 Proposed schematic illustration of the interaction between the cytoskeletal membrane-associated protein spectrin and actin, ankyrin, the ion channel (band 3), and the membrane protein glycophorin.

association with a number of other membrane proteins and other cytoskeletal elements. The spectrin heterodimers self-associate head to head to form 200-nm-long tetramers. Tetramers are then linked together by short actin filaments or another protein called Band 4.1, which binds spectrin, actin, and glycophorin, a transmembrane protein that has been identified in red blood cells.

Ankyrin is an important cytoskeletal protein that links spectrin to the cell membrane (Fig. 5.4). Ankyrin binds spectrin to a transmembrane protein called Band 3 protein, which is so named because of its position on polyacrylamide gel electrophoresis. Band 3 is an ion transport protein at the cell membrane that controls intracellular pH. Ankyrin has also been shown to be alternatively spliced in a tissue-specific manner similar to the COOH-terminus of dystrophin (40,41). It is tempting to speculate that this may impart a functional diversity in different cell types for the ankyrin molecule. Defects in the ankyrin protein have been shown to cause an abnormality in the cytoskeletal resulting in hereditary spherocytosis. This is a hereditary hemolytic anemia characterized by an abnormal spherical membrane of the red blood cells with characteristic increased fragility of the membrane to osmotic or shear force lysis.

The spectrin/ankyrin is linked to other important membrane proteins (36), including the α subunit of Na^+/K^+-ATPase as well as voltage-dependent sodium channels. β-spectrin is also believed to bind the acetylcholine receptor, and it is thought that this binding is sensitive to phosphorylation of the β-chain at multiple phosphorylation sites occurring at the COOH-terminus. Spectrin is also found to bind calmodulin, an important regulator of intracellular calcium metabolism. Association of spectrin with calmodulin has been shown to decrease spectrin-ankyrin interactions in vitro. In neuronal tissue, spectrin has been associated with synaptic junctions and the nodes of Ranvier and has also been found to be associated with the glutamine receptor, the nicotine receptor, and integrin molecules such as N-CAM, a neural cell adhesion molecule. Muscle tissue spectrin appears to be localized to the cell membrane as well as the sarcomeric Z-line, which may provide a link of the contractile apparatus with the membranous cytoskeleton. Multiple protein isoforms of spectrin have been identified, and the finding of cardiac-specific isoforms of spectrin may be important for an adaptive role of the cytoskeleton in cardiac myocyte (42).

Dystrophin

Mutations in the gene encoding for the cytoskeletal protein dystrophin are responsible for the skeletal and cardiac disease seen in Duchenne and Becker muscular dystrophies (43). The discovery of the dystrophin gene, and the subsequent identification of its RNA transcript and its protein, illustrate the power of molecular biology techniques in the study of inherited genetic diseases that influence the heart. In this case the molecular basis of muscle and cardiac dysfunction and necrosis in Duchenne and Becker muscular dystrophies was not understood until genetic studies mapping the disease to the short arm of the X chromosome targeted cloning of human DNA from this region of the chromosome and led to the discovery of this gene. The identification of the RNA product of this gene provided the information needed to predict the peptide sequence of the protein (44), develop antibodies, and identify the defect as abnormal expression of the cytoskeletal protein at the cytoplasmic surface of the cell membrane (45).

Mutations in the dystrophin gene often cause low-level production of a nonfunctional protein or complete absence of the dystrophin protein in the cells where it is normally expressed. Dystrophin is normally localized to the cell membrane of cardiac and skeletal muscle and smooth muscle, and to postsynaptic neurological junctions in the brain (46). Recent studies have also shown it to be an integral component of the membrane in human cardiac Purkinje fibers (47). As one might expect, the clinical phenotype of patients with Duchenne and Becker muscular dystrophies includes skeletal myopathy, cardiomyopathy, mental retardation, and conduction defects in the heart. While the function of dystrophin is still not completely understood, physiological studies have shown that dystrophic muscle has altered intracellular calcium metabolism. Elevated intracellular levels of calcium from skeletal and cardiac muscle from these patients appear to be related to an increase in calcium leak channels expressed at the cell membrane as well as augmented intracellular release of calcium when stimulated by acetylcholine (48,49). These data have led to speculation that dystrophin is associated with molecules that modify calcium activity at the cell membrane.

Dystrophin purified from skeletal muscle has been shown to be in close association with a number of integral membrane glycoproteins at the cell membrane (35) (Fig. 5.5). When dystrophin is not expressed in

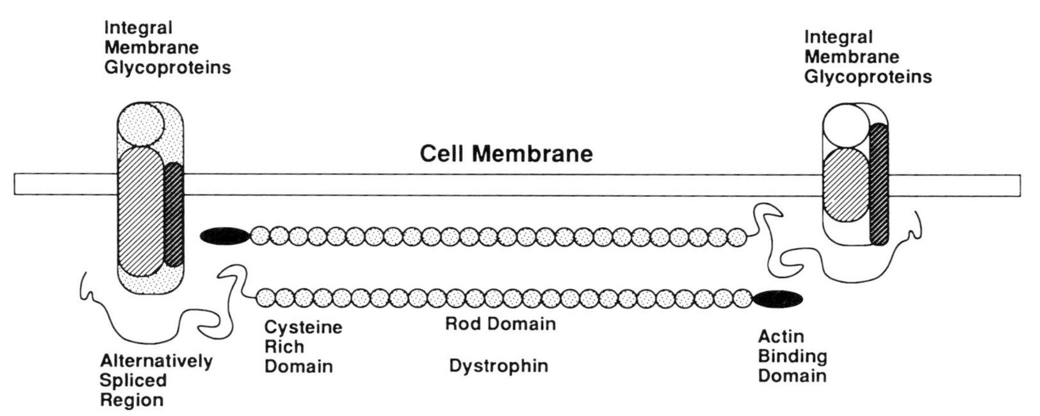

Figure 5.5 Schematic representation of the association of dystrophin with integral membrane proteins at the cytoplasmic surface of the cell membrane. The actin-binding, cysteine-rich, and rod-like domain structures are based on similarities between dystrophin and the other cytoskeletal proteins discussed (α-actinin and spectrin).

dystrophic muscle, some of the glycoproteins are also absent, indicating that dystrophin may be required to anchor this molecular complex of the cell membrane. It has been speculated that the function of dystrophin is imparted through its association with these molecules.

Dystrophin mRNA is alternatively spliced, encoding for multiple COOH-terminal isoforms of the dystrophin protein (50). These isoforms appear to be distributed in a tissue-specific pattern, which may confer a functional diversity for dystrophin in the tissues where it is expressed (38). The COOH-terminal region of the dystrophin molecule appears to contain the critical functional domains required for normal function. Several studies have shown that mutations that result in deletions of the COOH-terminal region of the dystrophin molecule lead to a severe disease phenotype, whereas mutations in the central domain of the molecule more often give a mild phenotype of the disease (43).

Dystrophin, spectrin, and α-actin also share a central rodlike domain of the molecule made up of repeating units. This rodlike domain is believed to be the structural portion of the molecule, giving strength and shape to the molecule. The rod domain is an α-helical structure that allows these molecules to self-associate as an antiparellel dimer, which can then polymerize at the cell membrane via actin and membrane protein linkages.

The dystrophin molecule has been shown to be homologous to other cytoskeletal proteins, including spectrin and α-actin, and all three of these molecules share an NH_2-terminal domain that can bind actin to form a latticelike network of cytoskeletal elements at the cytoplasmic surface of the cell membrane. The actin-binding domain may also link the cytoskeleton to actin stress fibers or actin associated with the sarcomeric Z-line. In muscle tissues, actin, intermediate filaments including desmin and vimetin, as well as the microtubules are likely involved in linking the cytoplasmic structural and contractile architecture to the cell membrane.

Summary

The cardiac sarcomere consists of a limited number of contractile proteins, each of which can exist as multiple isoforms, encoded by separate genes or resulting from alternative splicing. The particular isoforms expressed in myocardial cell are regulated throughout development and are often distinct in the atria versus ventricles. The expression of a fetal phenotype in overloaded and diseased myocardium is perhaps best demonstrated by isoform transitions that occur in the sarcomere. The specific molecular mechanisms that control and regulate these genes, and perhaps result in such pathologic transitions, are discussed in Chapter 6.

The cytoskeleton plays a critical role in the structural organization and integrity of the cell. Cytoskeletal proteins also interact with modulatory and structural elements and provide specific anchor sites for a number of important receptor molecules at the membrane and for molecules within the cell that modulate cell function and structure. Genetic diseases such as Duchenne muscular dystrophy and hereditary spherocytosis illustrate the importance of cytoskeletal proteins and provide insight into the nature of their interactions and functions.

References

1. Darnell J, Lodish H, Baltimore D (eds): Eukaryotic chromosomes and genes: Molecular anatomy. *Molecular Cell Biology*. New York: WH Freeman and Company, 1990, pp 347–390.
2. Darnell J, Lodish, Baltimore D (eds): Actin, myosin, and intermediate filaments: Cells movements and cell shape. *Molecular Cell Biology*. New York: WH Freeman and Company, 1990, pp 859–902.
3. Darnell J, Lodish H, Baltimore D (eds): Cell-to-cell signaling: Hormones and receptors. *Molecular Cell Biology*. New York: Freeman and Company, 1990, pp 709–762.
4. Alberts B, Bray D, Lewis J, et al (eds): Control of gene expression. *Molecular Biology of the Cell*. New York: Garland Publishing, Inc, 1983, pp 550–610.
5. Hirzel HO, Tuchschmid DR, Schneider J, et al: Relationship between myosin isoenzyme composition, hemodynamics, and myocardial structure in various forms of human cardiac hypertrophy. *Circ Res* 1985;57:729–740.

6. Imamura SI, Matsuoka R, Hiratsuka E, et al: Local response to cardiac overload on myosin heavy chain gene expression and isozyme transition. *Circ Res* 1990;66:1067–1073.
7. Nadal-Ginard B, Mahdavi F: Molecular basis of cardiac performance: Plasticity of the myocardium generated through protein isoform switches. *J Cin Invest* 1989;84:1693–1700.
8. Parker TG, Schneider MD: Growth factors, proto-oncogenes, and plasticity of the cardiac phenotype. *Annu Rev Physiol* 1991;53:173–200.
9. Parker G, Chow K-L, Schwartz RJ, Schneider MD: Differential regulation of skeletal α-actin transcription in cardiac muscle by two fibroblast growth factors. *Proc Natl Acad Sci USA* 1990a;87:7066–7070.
10. Parker TG, Packer SE, Schneider MD: Peptide growth factors can provoke "fetal" contractile protein gene expression in rat cardiac myocytes. *J Clin Invest* 1990;85:507–514.
11. Schneider MD, Roberts R, Parker TG: Modulation of cardiac genes by mechanical stress: The oncogene signalling hypothesis. *Mol Biol Med* 1991;8:167–183.
12. Black FM, Packer SE, Parker TG, et al: The vascular smooth muscle α-actin gene is reactivated during cardiac hypertrophy provoked by load. *J Clin Invest* 1991;88:1581–1588.
13. Schiliwa M: *The Cytoskeleton.* New York: Springer-Verlag, 1986.
14. Wang K: Sarcomere-associated cytoskeletal lattices in striated muscle. Review and hypothesis. In Shay JW (ed): *Cell and Muscle Motility*, New York: Plenum Press, 1985, pp 315–369.
15. Bray D, Heath J, Moss D: The membrane-associated "cortex" of animal cells: Its structure and mechanical properties. *J Cell Sci* 1986;4(suppl):71–88.
16. Korn ED, Carlier M-F, Pantaloni D: Actin polymerization and ATP hydrolysis. *Science* 1987;238:638–644.
17. Sato M, Schwartz WH, Pollard TD: Dependence of the mechanical properties of actin/α-actinin gels on deformation rate. *Nature* 1987;325:828–830.
18. Matsudaira P, Janmey P: Pieces in the actin-severing protein puzzle. *Cell* 1988;54:139–140.
19. Tilney LG: Interactions between actin filaments and membranes give spatial organization to cells. In Mcintosh JR, Satir BH (eds): *Modern Cell Biology*, vol 2: *Spatial Organization of Eukaryotic Cells.* New York: Alan R. Liss, 1983; pp 163–199.
20. Lee C, Chen LB: Dynamic behavior of endoplasmic reticulum in living cells. *Cell* 1988;54:37–46.
21. Allan VJ, Dreis TE: A microtubule-binding protein associated with membranes of the Golgi apparatus. *J Cell Biol* 1986;103:2229–2239.
22. Dustin P: *Microtubules*, 2nd ed. New York: Springer-Verlag, 1984; pp 127–164.
23. Sullivan KF: Structure and utilization of tubulin isotypes. *Annu Rev Cell Biol* 1988;4:687–716.
24. Amos LA, Baker TS: The three dimensional structure of tubulin protofilaments. *Nature* 1979;279:607–612.
25. Salmon D, McKeel M, Hays T: Rapid rate of tubulin dissociation from microtubules in the mitotic spindle *in vivo* measured by blocking polymerization with colchicine. *J Cell Biol* 1984;99:1066–1075.

26. Maruta H, Greer K, Rosenbaum JL: The acetylation of α-tubulin and its relationship to the assembly and disassembly of microtubules. *J Cell Biol* 1986;103:571–579.
27. Bara HS, Arce CA, Argarana CE: Posttranslational tyrosination detyrosination of tubulin. *Mol Neurobiol* 1988;2:133–153.
28. Olmstead JB: Microtubule-associated proteins. *Annu Rev Cell Biol* 1986;2:421–457.
29. Johnson KA: Pathway of the microtubule-dynein ATPase and structure of dynein: A comparison with actomyosin. *Annu Rev Biophys Biophys Chem* 1985;14:161–188.
30. Geisler N, Weber K: Phosphorylation of desmin *in vitro* inhibits formation of intermediate filaments: Identification of three kinase A sites in the aminoterminal head domain. *EMBO J* 1988;7:15–20.
31. Steinert PM, Roop DR: Molecular and cellular biology of intermediate filaments. *Annu Rev Biochem* 1988;57:593–626.
32. Quinlan RA, et al: Characterization of dimer subunits of intermediate filament proteins. *J Mol Biol* 1986;192:337–349.
33. Goldman R, et al: Intermediate filaments: Possible functions as cytoskeletal connecting links between the nucleus and the cell surface. *Ann NY Acad Sci* 1985;455:1–17.
34. Coleman TR, Fishkind DJ, Mooseker MS, Morrow JS: Functional diversity among spectrin isoforms. *Cell Motil Cytoskeleton* 1989;12:225–247.
35. Erasti JM, Ohlendieck K, Kahl S, et al: Deficiency of a glycoprotein component of the dystrophin complex in dystrophic muscle. *Nature* 1990;345:315–319.
36. Nomura H, Kizawa K: Histopathological study of the conduction system of the heart in Duchenne progressive muscular dystrophy. *Acta Pathol* 1982;32:1027.
37. Katiyar BC, Soman PN, Misra S, Chaterji AM: Congestive cardiomyopathy in a family of Becker's X-linked muscular dystrophy. *Postgrad Med J* 1977;53:12.
38. Bies R, Phelps SF, Cortez MD, et al: Human and murine dystrophin mRNA transcripts are differentially expressed during skeletal muscle, heart, and brain development. *Nucleic Acids Res* (submitted).
39. Alberts BE, Bray D, Lewis J, et al (eds): The cytoskeleton. *Molecular Biology of the Cell*. New York: Garland Publishing, Inc, 1983, pp 613–680.
40. Lux SE, John KM, Bennett V: Analysis of cDNA for human erythrocyte ankyrin indicates a repeated structure with homology to tissue-differentiation and cell cycle proteins. *Nature* 1990;344:36–42.
41. Lambert S, Yu H, Prchal JT, et al: cDNA sequence for human eryhtrocyte ankyrin. *Proc Natl Acad Sci USA* 1990;87:1730–1734.
42. Vybiral T, Williams JK, Winkelmann JC et al: Human cardiac and skeletal muscle spectrins: Differential expression and localization. *Cell Motil Cytoskeleton* 1992;21:291–304.
43. Koenig M, Beggs AH, Moyer M, et al: The molecular basis for Duchenne versus Becker muscular dystrophy: Correlation of severity with type of deletion. *Am J Hum Genet* 1988;45:498–506.
44. Koenig M, Monaco AP, Kunkel LM: The complete sequence of dystrophin predicts a rod shaped cytoskeletal protein. *Cell* 1988;53:219–288.
45. Hoffman EP, Brown RH, Kunkel LM: Dystrophin: The protein product of the Duchenne muscular dystrophy locus. *Cell* 1987;51:919–928.

46. Lidov HGW, Byers TJ, Walkins SC, Kunkel LM: Localization of dystrophin to postsynaptic regions of central nervous system cortical neurons. *Nature* 1990;348:725–728.
47. Bies RD, Friedman D, Perryman MB, et al: Expression and localization of dystrophin in human cardiac Purkinje fibers. *Circulation* 1992;86.
48. Fong P, Turner PR, Denetclaw WF, Steinhadt RA: Increased activity of calcium leak channels in myotubes of Duchenne human and mdx mouse origin. *Science* 1990;250:673–676.
49. Moneini T, Ghigo D, Doriguezzi C, et al: Free cytoplasmic Ca^{++} at rest and after cholinergic stimulus is increased in cultured muscle cells from Duchenne muscular dystrophy patients. *Neurology* 1988;38:476–480.
50. Feener CA, Koenig M, Kunkel LM: Alternative splicing of human dystrophin mRNA generates isoforms at the carboxy terminus. *Nature* 1989;338:509–511.

Chapter **6**

Molecular Biology of Cardiac Growth and Hypertrophy

The failure of cardiac muscle to regenerate following myocardial infarction and, conversely, the excessive growth associated with hemodynamic load continue to impose fundamental biological limits to clinical cardiology. The study of normal cardiac growth, development, and hypertrophy has provided a challenging yet fruitful field for the application of the tools of molecular genetics, providing information unobtainable by physiological or biochemical techniques alone. The structural and functional alterations in overloaded myocardium have provided a useful paradigm for the molecular biology of cardiac muscle. This chapter provides a review of recent advances in understanding the molecular events occurring in this process, including: 1) the spectrum of alterations of cardiac gene expression during pressure overload hypertrophy or myocardial disease and their impact on cardiac performance, 2) molecular signals induced by hemodynamic stress that might provoke growth and genetic alterations, 3) the role of peptide growth factors in regulating the cardiac phenotype, 4) cellular oncogenes and signal transduction, and 5) future directions in research.

Plasticity of Cardiac Gene Expression

Hypertrophic Induction of "Fetal" Gene Expression

Cardiac development and postnatal maturation entail both the neonatal loss of cardiac myocytes' proliferative capacity and sequential changes in cardiac muscle-specific genes (1). The prototypical alterations involve isoforms of the contractile protein genes, resulting from selective transcription of individual genes within multigene families (e.g., α-myosin heavy chain [α-MHC] and cardiac α-actin genes in the adult, β-myosin heavy chain [β-MHC] and skeletal α-actin genes in the fetus) or from alternative splicing of mRNA. In rodent models, pressure overload induced by aortic constriction provokes increased cardiac mass associated with cellular hypertrophy, accompanied by contractile protein gene expression characteristic of the embryonic heart. Thus, the decrease in myosin ATPase activity in overloaded hearts results from a transition from adult (V1, aa) to fetal (V3, bb) myosin heavy chains, which possess intrinsically different actin-activated properties (2). In addition, skeletal α-actin, β-tropomyosin, and atrial myosin light chain genes, all normally found only in the fetal ventricle, are reexpressed. This ensemble of fetal contractile proteins results in the formation of functionally distinct sarcomeres in the setting of cardiac hypertrophy (3) (Table 6.1).

The recapitulation of a fetal phenotype is more generalized, as evidenced by the reappearance of embryonic isoforms of myocardial metabolic enzymes (brain creatine kinase and lactate dehydrogenase) and the plasma membrane Na^{+}/K^{+}-ATPase (a_3 isoform) (4–8). In the absence of isoform switching, inhibition of the sarcoplasmic reticulum Ca^{2+}-ATPase and up-regulation of the atrial natriuretic factor genes also restore gene expression to that seen in embryonic hearts. This program of fetal gene expression is seen in other forms of experimental hypertrophy, including spontaneously hypertensive rats and Syrian hamster cardiomyopathy, and following myocardial infarction. Thus acute load is not mandatory for these genetic alterations. In contrast β-MHC is not provoked by the hypertrophy associated with exercise or thyroid hormone administration, and the specific subsets of genes modulated are dependent on both age and severity or duration of load (9–11).

Modulation of Gene Expression in Human Myocardial Disease

Whether comparable genetic alterations have equivalent significance in humans has been questioned, primarily because the normal adult human ventricle predominantly expresses the β-MHC gene and skeletal α-actin gene expression in human hypertrophy is at best inconsistent (12). Despite such counter-examples, human myocardium is capable of modulating gene expression. Recently, sensitive techniques have demonstrated that the majority of cardiac fibers in the human ventricle do express the α-MHC gene, albeit at low levels. The number of fibers expressing the α-MHC gene is decreased in myocardial disease in proportion to the decrement in cardiac index. It has been postulated that such microheterogeneity of sarcomere assembly may be required for optimal cardiac performance. The β-MHC gene, associated with embryonic but not adult atria, is reexpressed in mitral valve disease in proportion to atrial diameter. Alternative isoforms of β-MHC have been reported in dilated human cardiomyopathy. The atrial myosin light chain gene is reexpressed in human ventricular disease, and the importance of regional load is demonstrated by the anomalous appearance of ventricular myosin light chain in the right and left atria in pulmonary hypertension and mitral regurgitation, respectively. As in rodent models, atrial natriuretic factor is induced and the sarcoplasmic reticulum Ca^{2+}-ATPase gene is repressed in end-stage diseased ventricles. Finally, the importance of al-

Table 6.1 Continuum of Responses of Cardiac Myocytes to Peptide Growth Factors, Resembling Fetal Gene Expression Activated by a Hemodynamic Load

			Rodent Ventricle[b,c]			Human Ventricle[b,c]			Human Atrium[b,c]	
Gene/Isoform	Agent[a]	Response[b]	Embryonic	Adult	Hypertrophy	Embryonic	Adult	Overload or Failure	Adult	Overload
MHC										
α	aFGF, bFGF, TGFβ1	↓	—	+ + +	↓	+	+	↓	+ + +	↓
	NE	↔								
	T3	↑								
β	aFGF, bFGF, TGFβ1	↑	+ + +	—	↑	+ + +	+ + +	↔	+	↑
	NE	↑								
	T3	↓								
α-Actin										
Cardiac	bFGF, TGFβ1	↔	+ +	+ + +	↔	n.a.[d]	+ + +	↔	+ + +	n.a.
	NE	↔								
	T3	↔								
	aFGF	↓								
Skeletal	bFGF, TGFβ1	↑	+ +	—, +	↑	n.a.	+	variable	+	n.a.
	NE	↑								
	T3	↔, ↑								
	aFGF	↓								
Smooth	aFGF, bFGF, TGFβ1	↑	+ +	—	↑[e]	n.a.				
ANF	aFGF, bFGF, TGFβ1	↑	+ + +	—	↑	+ + +	—	↑	+ + +	↑
Slow/cardiac Ca^{2+}/ATPase	aFGF, bFGF, TGFβ1	↓	+	+ + +	↓	+	+ + +	↓		

[a] aFGF, acidic fibroblast growth factor; bFGF, basic fibroblast growth factor; NE, norepinephrine; T3, thyroid hormone; TGFβ1, type β1 transforming growth factor.
[b] ↑, up-regulation; ↓, down-regulation; ↔, little or no change with trophic signals or myocardial disease.
[c] Relative (+, + +, + + +) or absent (—) expression in normal myocardium. For brevity the sequence of gene induction during cardiac myogenesis is omitted.
[d] Not available or inconclusive.
[e] Data from Black et al. (26).

tered gene expression in human myocardial disease and hypertrophy has been demonstrated by the recent report of a mutation in the β-MHC gene and formation of a novel α/β-MHC fusion gene in two families with hypertrophic cardiomyopathy (13,14).

Does Altered Gene Expression Reflect Adaptation or Shared Regulatory Signals?

An attractive hypothesis is that plasticity of gene expression, both in human disease and in experimental models, allows cardiac muscle to alter its protein composition and potentially provides the basis for long-term adaptive changes to the hemodynamic burden, which contrasts the beat-to-beat fluctuations attributable to fiber length and the short-term biochemical control of excitation-contraction coupling. Indeed, the lower energetic cost of work and reduced velocity of shortening associated with the rate of actin-activated adenosine triphosphate (ATP) hydrolysis of β-MHC rather than α-MHC is presumed to be advantageous in the face of the high oxygen demand entailed in cardiac hypertrophy. Such adaptive plasticity could also be invoked with the presumed salutary effects of atrial natriuretic factor in volume overload. However, functional consequences of skeletal α-actin expression have not been demonstrated, and the reduced expression of the sarcoplasmic reticulum Ca^{2+}-ATPase conceivably would augment diastolic Ca^{2+} in the cytosol, impairing ventricular relaxation, although perhaps enhancing systolic function. Thus, the genetic alterations provoked by pressure overload may be advantageous, indifferent, or deleterious, and could reflect shared regulatory mechanisms rather than adaptation or homeostasis.

Initiating Signals for Pressure Overload Hypertrophy

Role of Circulating Neurohumoral Factors

The specific molecular signals that convert hemodynamic load into an altered cardiac phenotype remain to be elucidated. Although thyroid hormone is the canonical regulator of MHC gene expression, thyroid hormone levels are not altered during pressure overload and this agonist is incapable of modulating the spectrum of genetic alterations discussed above. Whereas catecholamine levels are elevated and norepinephrine can induce hypertrophy, β-MHC, and skeletal α-actin in cultured cardiac cells, the relative absence of atrial and right ventricular hypertrophy in aortic coarctation and the selective growth of right ventricular myocytes in pulmonary banding argue strongly against a role for circulating factors in initiating cardiac hypertrophy and the associated fetal phenotype. The potential role of alternative local autocrine or paracrine mechanisms remains a viable hypothesis (Table 6.1).

Role of Mechanical Stress Itself

That perturbations induced by load are local argues for a contribution from mechanical stress or cellular deformation itself (15,16). This hypothesis is supported by the demonstration that obstruction to left ventricular outflow results in hypertrophy of the ventricular wall but not of unloaded papillary muscles. Passive stretch of isolated ventricular myocytes on distensible membranes induces protein and RNA synthesis, β-MHC, skeletal α-actin, and atrial myosin light chain. Although the role of putative stretch-activated ion channels in this process remains speculative (17–20), four intracellular signaling pathways are stimulated by stretch alone, including cyclic adenosine monophosphate, phosphatidylinositol hydrolysis, intracellular calcium, and the oncogene-encoded transcription factors described later in this chapter.

A straightforward mechanical model of signal transduction is, however, contradicted by several lines of evidence in which facets of hypertrophy are dissociated from load per se. Cardiac hypertrophy precedes hypertension in spontaneously hypertensive rats, and β-blocking agents, which prevent increased cardiac mass, do not inhibit β-MHC gene expression in this preparation. Similarly, in humans, alternate antihypertensive regimen have differential effects on regression of cardiac hypertrophy despite comparable control of blood pressure. Finally, temporal and spatial discrepancies in the appearance of fetal gene expression following aortic coarctation cannot be accounted for by alterations in load alone. Thus, molecular signals other than load may modulate cardiac growth and gene expression, potentially by local autocrine or paracrine mechanisms.

Role of Peptide Growth Factors in Cardiac Hypertrophy

The observation that crude extacts from hypertrophied hearts could provoke hypertrophy when infused in vivo provided the first evidence that trophic substances are locally produced in the myocardium, although the specific peptides have yet to be characterized. An increasing number of specific polypeptides that influence cellular growth have been identified from a variety of studies of cellular proliferation, transformation, and angiogenesis. Such studies have also defined the role of these peptides in modulating differentiation and gene expression in a lineage- and developmentally specific manner. In cardiac muscle, the heparin binding or fibroblast growth factors (FGFs) and type β transforming growth factors (TGFβ) are of particular interest given their expression, developmental regulation, and induction during myocardial disease.

Expression and Regulation through Development

Acidic and basic FGF are distinct but highly related peptides synthesized in myocardium, in part by myocytes themselves, and stored in the extracellular matrix, such that cardiac muscle is a particularly abundant source of acidic FGF. TGFβ1 is also expressed by cardiac myocytes and, like basic FGF, is expressed at higher levels in the atria than the ventricles, in parallel with the differential capacity of those chambers for DNA synthesis. A variety of related peptides (TGFβ2–5) are also expressed in the hearts of several species. Other trophic peptides found in myocardium include a TGFβ-related peptide, insulinlike growth factors I and II, and at least three nerve growth factors. Cardiac fibroblasts also produce distinct trophic substances.

TGFβ is found at high levels in the endocardial cushions and valve primordia in embryonic mice, and functional studies demonstrate that both TGFβ and FGFs may contribute to formation of valve mesenchyme from embryonic epithelium. Similarly, these and related peptides are important in the induction of cardiac muscle itself in the early embryo at the onset of gastrulation, indicating their likely critical role in regulating cardiac growth and gene expression. Basic FGF and TGFβ1–4 are down-regulated in adult myocardium relative to fetal levels, further suggesting a possible link to the loss of cardiac myocytes' proliferative capacity and the associated changes in gene expression.

Expression in Cardiac Disease

Increased expression of peptide growth factors has been demonstrated with myocardial ischemia, infarction, and altered load. Ischemia in the pig results in up-regulation of TGFβ in ventricular myocytes and acidic FGF in coronary vessels. Both TGFβ1 and basic FGF are lost from infarcted cardiac muscle in the rat but are up-regulated in the surviving surrounding myocytes (21). Thus, these peptides may play a role in infarct healing, neovascularization, and compensatory hypertrophy. Recently, the early augmented expression of basic FGF and insulinlike growth factor I in the cardiac myocytes of pressure-overloaded hearts has been demonstrated. Regulation of acidic FGF in this setting has not been studied, but acidic FGF is up-regulated tenfold in skeletal muscle by skeletal conditioning. These observational studies furnish intriguing hypothesis as to the role of peptide growth factors that demand functional testing of the effects of these peptides on growth and gene expression in cardiac muscle.

Skeletal Muscle as a Model of Growth Factor Effects

Historically, skeletal muscle has served as a favorable model for the effects of peptide growth factors on striated myogenesis, in large part because of the mutually exclusive relationship between cellular proliferation and the expression of muscle-specific genes. That is, proliferating skeletal myoblasts do not express proteins that are characteristic of the muscle phenotype. Furthermore, as is not the case for cardiac muscle, well-characterized cell lines exist that exhibit differentiated (i.e., muscle-specific, including myosin, actin, etc.) gene expression only on withdrawal of serum from the culture medium, permitting identification of specific molecules regulating skeletal myogenesis. Finally, the introduction and expression of foreign genes in such cell lines enables analysis of DNA regulatory sequences required for muscle gene expression, and the role of specific gene products in muscle growth and differentiation.

Both basic and acidic FGF result in proliferation of skeletal myoblasts while concomitantly inhibiting myogenic differentiation. In contrast, TGFβ1 similarly blocks muscle-specific gene expression but has no mitogenic activity, suggesting the potential for independent control of growth and differentiation. The single extracellular ligand, TGFβ1, in fact, can reversibly block the expression of an entire program of gene expression, entailing not only sarcomeric proteins but also muscle creatine kinase and the sarcolemmal Ca^{2+} and Na^{+} channels. However, the effects of growth factors are dependent on the particular stage of development of the treated cultures. TGFβ1 and basic FGF down-regulate muscle-specific genes in differentiated skeletal muscle prior to formation of myotubes but have little effect on the muscle phenotype of terminally differentiated, fused cultures despite the persistence of functional cell surface receptors. The uniform, yet developmental stage–dependent, suppression of the entire skeletal muscle phenotype by TGFβ and FGFs is likely dependent on their interaction with "master regulatory" genes, such as myo-D1. Myo-D1 is a member of a family of "determination" genes that are turned on early in development in cells destined to form skeletal muscle (22). The protein product of myo-D1 then, in turn, initiates the transcription of other muscle-specific genes, including those for sarcomeric proteins. TGFβ1 can inhibit the expression of myo-D1, and other peptide growth factors, including basic FGF, can modulate the activity of the myo-D1 protein, proving hierarchical mechanisms for the inhibition of muscle-specific genes.

Peptide Growth Factors Provoke Fetal Gene Expression in Cultured Cardiac Muscle

Although the well-delineated function of peptide growth factors in muscle-specific gene expression in skeletal muscle suggests a high likelihood that cardiac myocytes are also targets for their action, important differences exist between skeletal and cardiac myogenesis (Fig. 6.1). Unlike skeletal muscle, cardiac myocytes synthesize muscle proteins while still undergoing proliferative growth in the embryo. Cardiac muscle loses proliferative potential in the early newborn period (while retaining some

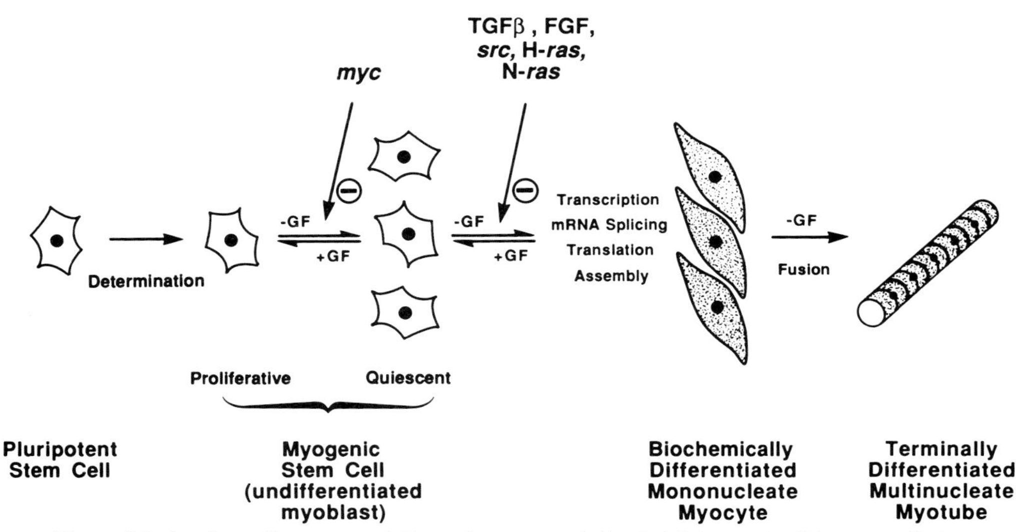

Figure 6.1 A schematic representation of myogenesis in skeletal muscle. It is generally believed that, once myotubular formation occurs in skeletal muscle, differentiation is irreversible.

capacity for DNA synthesis, resulting in a normally tetraploid or binucleate state), and subsequent adaptive growth is by cellular enlargement only. A viral oncogene, SV40 T antigen, which induces proliferation in both skeletal and cardiac muscle, is permissive for differentiated properties only in the cardiac lineage. The basis for these differences is unknown but, of interest, the myo-D1–like family of "determination" genes is not appreciably expressed in the heart.

Cardiac myocytes in primary culture have enabled molecular analysis of cardiac growth and hypertrophy in the absence of confounding hemodynamic and systemic variables. Early evidence that peptide growth factors could affect the cardiac phenotype was the observation that cardiac muscle cultures maintained in crude serum, which contains various peptide mitogens, progressively lost their differentiated state. In contrast, culture of neonatal rat ventricular myocytes in serum-free media maintained, or even advanced, differentiation. Reexposure of these cultures to serum after successively longer periods of serum withdrawal produces proliferation, DNA synthesis alone, or neither growth response in keeping with normal developmental changes. Serum-induced proliferation is associated with down-regulation of α-actin expression, but selectively induces skeletal α-actin in older cultures.

The specific response of cardiac myocytes to TGFβ1 and basic and acidic FGF has also been studied in cultured neonatal ventricular myocytes maintained in serum-free medium. Both TGFβ1 and basic FGF produce a pattern of contractile protein gene expression strongly resembling that seen with pressure overload hypertrophy in vivo (Fig. 6.2) (23–26). Each ligand inhibits adult α-MHC by two thirds, while inducing the expression of the fetal β-MHC fourfold. In addition, each stimulates fetal skeletal α-actin expression with little effect on the cardiac α-actin gene. TGFβ1 produces these effects in the absence of cellular growth, whereas basic FGF stimulates total protein production in the cultures in the absence of proliferation, as would be seen with cellular hypertrophy. Despite also inducing β-MHC, acidic FGF does provoke proliferation in the serum-free cardiac cultures, which, as with the serum-exposed cultures, was associated with the down-regulation of both cardiac and skeletal α-actin. The specificity of the effects on α-actin gene expression has been confirmed by the ability of the growth factors to identically regulate the transcription of foreign actin genes transfected into cardiac muscle cells. In addition, all three peptides modulate a wide spectrum of other cardiac genes, including up-regulation in cardiac cultures of smooth muscle α-actin, a gene normally expressed only early in cardiac embryogenesis. Again, as seen with cardiac hypertrophy in vivo, these trophic peptides also induce atrial natriuretic factor expression while down-regulating the sarcoplasmic reticulum Ca^{2+}-ATPase.

Thus, the effects of FGFs and TGFβ on cardiac muscle demonstrate a complex and selective regulation of cardiac-specific genes that differs substantially from the uniform suppression seen in skeletal muscle. Moreover, the pattern of response not only provides insight into the previously defined actions of serum, but suggests an important role for peptide growth factors in the control of cardiac growth and modulation of gene expression in development and hypertrophy. The generalized fetal phenotype produced by TGFβ1 and basic FGF specifically implies the potential for a paracrine or autocrine role of growth factors in pressure overload hypertrophy. Other peptides with trophic and regulatory effects on cardiac muscle, including insulinlike growth factors, angiotensin, endothelin, and a heparin binding peptide produced by cardiac mesenchymal cells, are undergoing investigation.

Cellular Oncogenes and Growth Factor Signal Transduction

Oncogene-Encoded Protein Transmission of Extracellular Growth Signals to the Nucleus

Peptide growth factors control cell growth and gene expression through a cascade of proteins encoded by cellular or proto-oncogenes. These cellular oncogenes were identified first as the normal homologues of transforming genes of retroviruses and subsequently as mammalian genes whose mutation could also transform cells in culture. Numerous

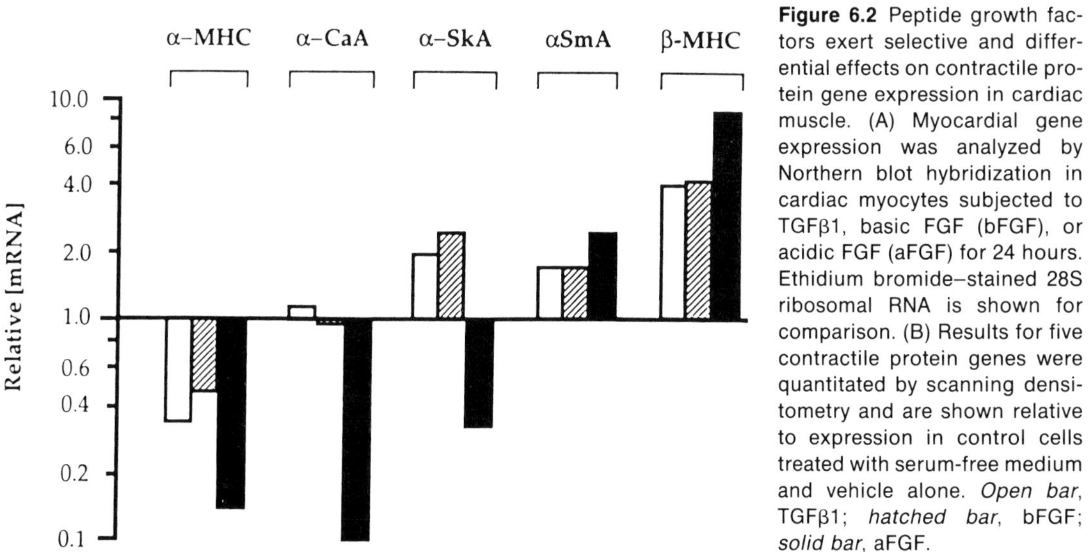

Figure 6.2 Peptide growth factors exert selective and differential effects on contractile protein gene expression in cardiac muscle. (A) Myocardial gene expression was analyzed by Northern blot hybridization in cardiac myocytes subjected to TGFβ1, basic FGF (bFGF), or acidic FGF (aFGF) for 24 hours. Ethidium bromide–stained 28S ribosomal RNA is shown for comparison. (B) Results for five contractile protein genes were quantitated by scanning densitometry and are shown relative to expression in control cells treated with serum-free medium and vehicle alone. *Open bar*, TGFβ1; *hatched bar*, bFGF; *solid bar*, aFGF.

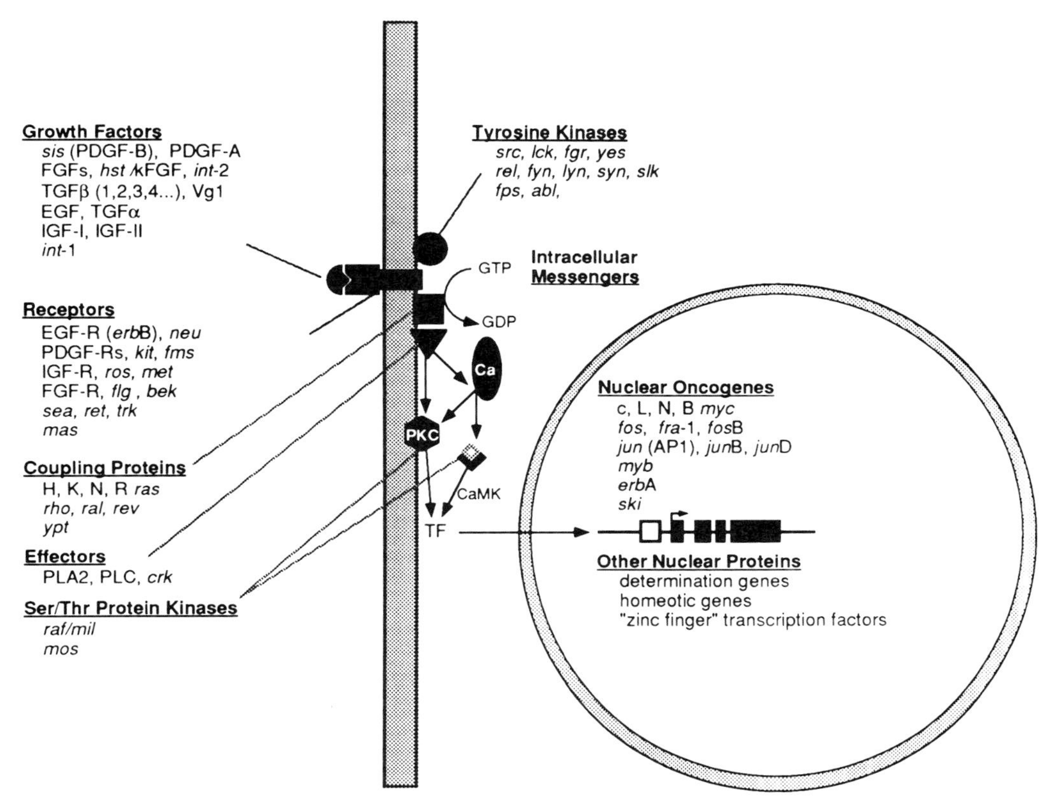

Figure 6.3 The many proteins with varied functions for which oncogenes are known to encode. It is clear from this diagram that oncogenes encode proteins that function as growth factors, receptors, coupling proteins, signaling proteins, and transcription factors.

cellular oncogenes are expressed in the heart, but their exact cellular localization remains unclear and their function has been inferred from simpler systems.

Oncogene products span the cell from membrane to nucleus and include growth factors themselves (Fig. 6.3). The proto-oncogene c-*sis* encodes for the B chain of platelet-derived growth factor, postulated to be critical in atherogenesis. Several oncogenes also encode for homologues of TGFβ and FGFs. A variety of cell surface receptors are critical in the cascade, and *erb*B encodes for the epidermal growth factor receptor. This gene may be down-regulated in human dilated cardiomyopathy, although the significance remains purely speculative.

Propagation of the signal from the cell surface to the nucleus is likely to be dependent on membrane-associated and cytoplasmic protein kinases and the phospholipases. Multiple genes encode for these peptides but, of interest, the oncogenes *crk* and *raf* are structurally similar to phospholipase C and protein kinase C, respectively. Protein kinase C can substitute for cell surface mitogens and, in cardiac muscle, can induce a co-transfected β-MHC gene. The cellular oncogene *ras* has been considered to have a pivotal role in signal transduction because *ras* proteins bind and hydrolyze guanosine triphosphate, akin to the G-proteins critical in β-adrenergic signal transduction. *Ras* proteins are abundant in cardiac muscle but appear to undergo minimal regulation in development or following aortic constriction. High-level *ras* expression prevents muscle-specific gene expression in skeletal myoblasts as seen with TGFβ1, but determination of its effect in cardiac muscle awaits study.

Nuclear Oncogene–Encoded Transcription Factors in Myocardium

Nuclear oncogene–encoded proteins—including c-*myc*, c-*fos*, and c-*jun*—are rapidly induced by growth factors and a variety of other trophic signals (26–28). These proteins are localized to the nucleus and are thought to couple growth factor signals to the transcriptional control of a variety of other genes. Thus, these proteins act as transcription factors and have received considerable attention in the context of cardiac growth and hypertrophy. Down-regulation of c-*myc* during cardiac development parallels the loss of myocyte proliferative capacity. Conversely, pressure overload hypertrophy, Syrian hamster cardiomyopathy, and hypertrophy in spontaneously hypertensive rats are all associated with the reexpression of c-*myc* in vivo. Norepinephrine induces c-*myc* and hypertrophy in cultured ventricular cells, whereas serum induced c-*myc* both in the presence and in the absence of proliferative growth. In agreement with the delay but not inhibition of differentiation in skeletal muscle cell lines, expression of a deregulated c-*myc* in the hearts of transgenic mice increases cardiac size by proliferation but does not immortalize cardiac myocytes in culture or alter the pattern of cardiac-specific gene expression. Taking these observations together, loss of c-*myc* is unlikely to account for loss of myocyte proliferative capacity in vivo and reinduction of c-*myc* precedes divergent phenotypic responses, suggesting a nonspecific relationship.

There is little or no steady state expression of c-*fos* in the developing rat heart, but this gene is rapidly up-regulated by load, aging, and diverse pharmacological trophic signals in vivo. In cell culture, c-*fos* is also induced by passive stretch, angiotensin, endothelin, and both basic and acidic FGF. Other members of the *fos* gene family include *fra*-1 and *fos*B. Unlike c-*myc*, c-*fos* is a confirmed transcription factor whose activation of gene expression is dependent on formation of a heterodimer with c-*jun* or *jun* homologue, which facilitates binding to regulatory DNA sequences. *Fos-jun* binds to the regulatory regions of the atrial natriuretic gene, which is activated in ventricular muscle during pressure overload hypertrophy. Three members of the *jun* family—c-*jun*, *jun*B, and *jun*D—

are expressed at basal levels in myocardium. In vivo, c-*jun* and *jun*B are induced by pressure overload and by acidic FGF in cultured myocytes. Preliminary evidence suggests a differential effect of basic and acidic FGF on c-*jun* induction. The complex genetic responses to growth factors in cardiac muscle could be dependent on the induction of different permutations of *fos-jun* heterodimers whose effects on gene transcription vary. For example, α-adrenergic agonists, but not β-agonists, induce the nucleoprotein EGR-1 and skeletal α-actin expression in cardiac muscle. A similar mechanism could account for the distinct effects of basic and acidic FGF.

The observation that induction of growth factor–inducible nuclear oncogenes precedes fetal gene expression in models of cardiac hypertrophy demands mechanistic testing of a cause-and-effect relationship. Molecular techniques of gene transfer or gene blocking can now be applied to the heart to determine the phenotypic consequences of overexpression or inhibition of oncogene-encoded proteins. Similarly, the DNA regulatory regions of cardiac-specific genes can be "dissected" to determine the interaction with nuclear oncoproteins and other transcription factors. Finally, the complex response of a wide variety of cardiac genes to single growth factors beckons a determined search for cardiac homologues of myo-D1, regulatory factors through which, analogous to skeletal muscle, these ligands might act.

Summary and Future Directions

Cardiac muscle is a genetically active tissue with a remarkable capacity to alter myocyte protein composition. Such plasticity is seen both in a wide variety of experimental models and in myocardial disease in humans. Cardiac hypertrophy induced by pressure overload has become a paradigm for the molecular analysis of the events regulating the cardiac phenotype because it is manifested not only by cellular growth but also by the reexpression of a generalized fetal program of gene expression in the subset of cardiac-specific genes studied to date. The induction of fetal genes is not unique to cardiac muscle, as demonstrated, for example, by the down-regulation of albumin and induction of α-fetoprotein during liver regeneration, and likely reflects a basic biological principle.

The molecular signals that translate increased aortic pressure into myocyte growth and altered gene expression are likely to be multifactorial (29). These include the effects of cellular deformation itself and the modulating effects of neurohumoral systems. The remarkable similarity in the pattern of gene expression (both muscle specific and nuclear oncogene) and growth seen in pressure overload hypertrophy in vivo and in response to peptide growth factors in vitro suggests at least shared intracellular transduction pathways. This provides an amenable model in which to study intracellular signal transduction, including second messengers and transcription factors. However, the regulation of these peptides through cardiac development and in response to pathological stimuli suggests the possibility that they play a role as autocrine or paracrine factors in inducing, sustaining, or modulating cardiac hypertrophy in vivo.

It should now be relatively straightforward to determine the degree of activation of peptide growth factors in cardiac hypertrophy in vivo and whether this activation correlates with fetal gene activation both temporally and topographically. Of perhaps greater interest will be tests of the role of exogenous peptide growth factors (or their antagonists) in modifying cardiac growth and gene expression in vivo in a variety of settings, including ischemia, infarction, and pressure overload. Perhaps these peptides offer the hope of obviating the present biological constraints of myocardial response to these stimuli.

References

1. Mulvagh SL, Roberts R, Schneider MD: Cellular oncogenes in cardiovascular disease. *J Mol Cell Cardiol* 1988;20:657–662.
2. Alpert NR, Mulieri LA, Litten RZ: Isoenzyme contribution to economy of contraction and relaxation in normal and hypertrophied hearts. In Jacob R, Bulch RW, Kissling G (eds): *Cardiac Adaptation to Hemodynamic Overload, Training and Stress*. Darmstadt, FRG: 1983, pp 147–157.
3. Schneider MD, Roberts R, Parker TG: Modulation of cardiac genes by mechanical stress: The oncogene signalling hypothesis. *Mol Biol Med* 1991;8:167–183.
4. Hammond GL, Nadal-Ginard B, Talner NS, Markert CL: Myocardial LDH isozyme distribution in the ischemic and hypoxic heart. *Circulation* 1979;53:637–643.
5. Herrera VL, Emmanuel JR, Ruiz-Opazo N, et al: Three differentially expressed Na,K-ATPase α subunit isoforms: Structural and functional implications. *J Cell Biol* 1987;105:1855–1865.
6. Komuro I, Kurabayashi M, Shibazaki Y, et al: Molecular cloning and characterization of a $Ca^{2+}+Mg^{2+}$-dependent adenosine triphosphatase from rat cardiac sarcoplasmic reticulum. *J Clin Invest* 1989;83:1102–1108.
7. Ueno H, Perryman MB, Roberts R, Schneider MD: Differentiation of cardiac myocytes following mitogen withdrawal exhibits three sequential states of the ventricular growth response. *J Cell Biol* 1988;107:1911–1918.
8. Schneider MD, Perryman MB, Payne PA, et al: Autonomous c-*myc* expression in transfected muscle cells does not prevent myogenic differentiation. *Mol Cell Biol* 1987;7:1973–1977.
9. Nadal-Ginard B, Mahdavi V: Molecular basis of cardiac performance: Plasticity of the myocardium generated through protein isoform switches. *J Clin Invest* 1989;84:1693–1700.
10. Gustafson TA, Markham BE, Bahl JJ, Morkin E: Thyroid hormone regulates

expression of a transfected α-myosin heavy chain fusion gene in fetal heart cells. *Proc Natl Acad Sci USA* 1987;84:3122–3126.
11. Hirzel HO, Tuchschmid CR, Schneider J, et al: Relationship between myosin isoenzyme composition, hemodynamics, and myocardial structure in various forms of human cardiac hypertrophy. *Circ Res* 1985;57:729–740.
12. Swynghedauw B: *Cardiac Hypertrophy and Failure*. London: John Libbey & Co. Ltd, 1990.
13. Tanigawa G, Jarco JA, Kass S, et al: A molecular basis for familial hypertrophic cardiomyopathy: A α/β cardiac myosin heavy chain gene hybrid gene. *Cell* 1990;62:991–998.
14. Hejtmancik JF, Brink PA, Towbin J, et al: Localization of the gene for familial hypertrophic cardiomyopathy to chromosome 14q1 in a diverse American population. *Circulation* 1991;83:1592–1597.
15. Hammond GL, Wieben E, Markert CL: Molecular signals for initiating protein synthesis in organ hypertrophy. *Proc Natl Acad Sci USA* 1979;76:2455–2459.
16. Mann DL, Kent RL, Cooper G: Load regulation of the properties of adult feline cardiocytes: Growth Induction by cellular deformation. *Circ Res* 1989;64:1079–1090.
17. Kaida T, Komuro I, Yazaki T: Increased protein synthesis and myosin isoform change in cultured cardiocytes by loading. *Circulation* 1988;78(suppl II):II-242.
18. Watson PA, Haneda T, Morgan HE: Effect of higher aortic pressure on ribosome formation and cAMP content in rat heart. *Am J Physiol* 1989;256:C1257–C1261.
19. von Harsdorf R, Lang RE, Fullerton M, Woodcock EA: Myocardial stretch stimulates phosphatidylinositol turnover. *Circ Res* 1989;65:494–501.
20. Marban E, Koretsune Y: Cell calcium, oncogenes and hypertrophy. *Hypertension* 1990;15:652–658.
21. Chiba M, Bazoberry F, Speir EH, et al: Role of basic fibroblast growth factor in angiogenesis: Healing and hypertrophy after rat myocardial infarction. *Circulation* 1989;80(suppl II):II-452.
22. Davis RL, Cheng PF, Lassar AB, Weintraub H: The MyoD DNA binding domain contains a recognition code for muscle-specific gene activation. *Cell* 1990;60:773–746.
23. Parker TG, Schneider MD: Growth factors, proto-oncogenes, and plasticity of the cardiac phenotype. *Annu Rev Physiol* 1991;53:173–200.
24. Parker G, Chow K-L, Schwartz RH, Schneider MD: Differential regulation of skeletal α-actin transcription in cardiac muscle by two fibroblast growth factors. *Proc Natl Acad Sci USA* 1990a;87:7066–7070.
25. Parker TG, Packer SE, Schneider MD: Peptide growth factors can provoke "fetal" contractile protein gene expression in rat cardiac myocytes. *J Clin Invest* 1990;85:507–514.
26. Black FM, Packer SE, Parker TG, et al: The vascular smooth muscle α-actin gene is reactivated during cardiac hypertrophy provoked by load. *J Clin Invest* 1991;88:1581–1588.
27. Payne PA, Olson EN, Hsiau P, et al: An activated c-Ha-*ras* allele blocks the induction of muscle-specific genes whose expression is contingent on mitogen withdrawal. *Proc Nat Acad Sci USA* 1987;84:8956–8960.

28. Schneider MD, Payne PA, Ueno H, et al: Dissociated expression of c-*myc* and a *fos*-related competence gene during cardiac myogenesis. *Mol Cell Biol* 1986;6:4140–4143.
29. Schneider MD, Parker TG: Molecular mechanisms of cardiac growth and hypertrophy: Myocardial growth factors and proto-oncogenes in development and disease. In Roberts R (ed): *Molecular Biology of the Cardiovascular System*. Hamden, CT: Blackwell Scientific Publications, Inc, 1992 (in press).

Chapter 7

Molecular Biology of Cardiac Ion Channels

Pediatric and adult cardiologists deal with a variety of electrophysiological abnormalities on a daily basis. Clinically this includes the spectrum from bradyarrhythmias to tachyarrhythmias, as well as drug interrelationships and their actions. Sudden death from arrhythmias occurring in patients with ischemic heart disease is a major cause of death in the Western world and accounts for 30% to 40% of patients who die prior to reaching the hospital. These arrhythmias occur as a result of abnormalities of the electrical system of the heart and various second messengers secondary to injury, primarily myocardial ischemia. Efforts to develop safe and effective antiarrhythmic pharmaceuticals have been blighted by the serious side effects and lack of specific antiarrhythmic effect of the various drugs. The results of these efforts to date, however successful or unsuccessful, were attained without knowledge of their target molecules. Most antiarrhythmias have been targeted to the sodium channel, yet the human cardiac sodium channel has yet to be cloned and sequenced. Results already indicate that sodium channels, like potassium and calcium channels, are somewhat tissue specific. The sequencing of the tissue subtypes of the different forms of each of these classes of channels should provide the opportunity to genetically design drugs that are specific for cardiac channels. In this chapter we hope to provide the necessary background information to appreciate the potential benefit to be derived from the application of such techniques in the near future.

Normal Cardiac Cellular Electrophysiology

The invention of the glass capillary microelectrode by Ling and Gerard in 1949 (1) enabled electrophysiologists to perform detailed studies of cardiac action potentials and the ionic events underlying cardiac electrical activity. Intracardiac electrical activity could then be studied by inserting a fine-tipped glass microelectrode of less than 0.5 nm into a single cardiac cell. A second electrode, remaining outside the cell, is used as a reference point. On advancing the microelectrode through the cell membrane, a large negative electrical potential is detected. Depending on the cell type studied, the inside of the cell may vary between −55 and −95 mV relative to the cell exterior. This negative electrical potential has been shown to be due to the distribution of ions across this cell membrane. The resting cardiac cell is a highly permeable to potassium and relatively impermeable to sodium, therefore behaving like a potassium electrode. As the concentration of extracellular potassium increases, the membrane potential decreases (i.e., moves toward 0 mV).

The determinants of the resting membrane potential can be divided into passive and active elements. The movement of ions down their electrochemical gradients constitutes the largest passive element, with potassium being the major ion determining this resting potential. Its intracellular concentration of 150 mmol/liter is nearly 40 times higher than the extracellular concentration (4 mmol/liter). This high intracellular concentration of potassium is maintained by the energy-requiring Na^+/K^+-ATPase pump, which brings the potassium into the cell and pumps sodium out of the cell. Sodium and calcium concentration gradients appear to have a minor role in influencing resting membrane potential. The cell membrane exhibits limited permeability to sodium during the steady state, with this sodium permeability being approximately 2% that of the potassium permeability. The contribution of sodium to the resting membrane potential is only apparent at a low external concentration of potassium (i.e., less than 1 mmol/liter). It is thought that sodium permeability is the result of the opening of a small fraction of voltage-dependent sodium channels (which are discussed later). Alteration of external calcium concentration has been shown to influence resting membrane potential, and it is believed that calcium may indirectly alter resting sodium permeability or that there may be a finite calcium permeability at rest. A second passive element contributing to resting membrane potential is the various fixed charge groups located on the membrane surface that give rise to a net negative charge to the cell interior. Phospholipids and glycoproteins are the major contributors of this group.

The electrochemical gradients established across the cell membrane are maintained by two active transport processes, the Na^+-K^+ pump and the Na^+-Ca^{2+} exchanger. Both of these transport processes contribute to the resting membrane potential by preventing the dissipation of the eletrochemical gradient direct contribution to the membrane potential by the net transport of charge. In the steady state, active sodium efflux must balance sodium influx. This is established through changes in the Na^+-K^+ pump activity in which three sodium ions are extruded in exchange for two potassium ions. This then produces a net outward current that hyperpolarizes the membrane. The Na^+-Ca^{2+} exchanger may either depolarize or hyperpolarize the cell membrane depending on the direction of exchange. One internal calcium ion is exchanged for two or three external sodium ions under normal conditions, and therefore this process will generate an inward depolarization current. When abnormally high intracellular concentrations exist, external calcium may be exchanged for internal sodium, thereby producing an outward depolarizing current.

When a stimulus of appropriate strength and duration depolarizes the cell to the level of threshold potential, there is an increase in sodium permeability that quickly depolarizes the cell membrane and produces a regenerative action potential. The transmembrane potential difference disappears and reverses itself as the cell's interior changes from −90 mV to +30 mV. The cell then repolarizes and the transmembrane potential returns to −90 mV. This is known as the cardiac action potential, which is the net result of a sequence of changes in the ionic permeabilities of the cell membrane, with different ions carrying charges into and out of the cell. In the normal heart two different types of action potentials are generated depending upon the type of cell studied. These action potentials are named for the rapidity of their upstroke and the initial part of the depolarization: "fast response" and "slow response" action potentials (Fig. 7.1). Atrial and ventricular myocardial cells and ventricular conduction cells normally have fast response action potentials, whereas the slow response action potential occurs in cells of the normal sinus node, the atrioventricular node, and the coronary sinus, and in some cells of the atrioventricular valves.

In the fast response action potential, the resting cell membrane is permeable to potassium ions but relatively impermeable to sodium ions, as previously mentioned. Microelectrode studies demonstrate a −90 mV charge inside with respect to the outside in these cells. When "fast responders" are stimulated, sodium permeability increases markedly and there is a rapid inflow of sodium that causes rapid depolarization such that the inside of the cell becomes positive with respect to the outside.

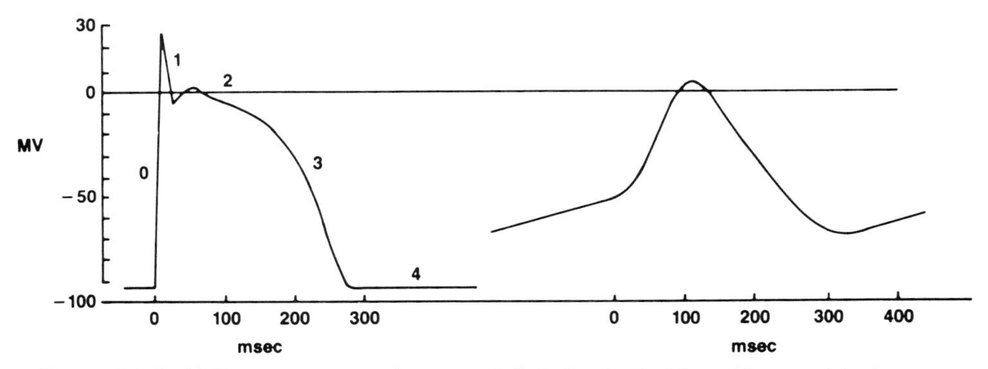

Figure 7.1 (Left) Fast response action potential similar to that found in a ventricular myocardial cell; the phases of the action potential are indicated by numbers 0 through 4. (Right) Slow response action potential similar to that found in a sinus node P cell.

Cardiac Action Potential

Cardiac Action Potential Phases

The cardiac action potential consists of the following five phases (Fig. 7.2).

Phase 0

This is the most rapid phase of the action potential and consists of rapid inflow of sodium. This phase is completed within several milliseconds. The rapidity with which phase 0 occurs is defined as the change in voltage with respect to time and is called the V_{max}; this correlates with the conduction velocity in the single cell. This upstroke (V_{max}) is much slower in the slow response action potential compared with that seen in the fast response action potential.

Phase 1

This early, rapid repolarization phase is primarily due to an outward repolarizing chloride current. At the same time the chloride current occurs, the inward sodium current is decreasing and the outward potassium current is beginning.

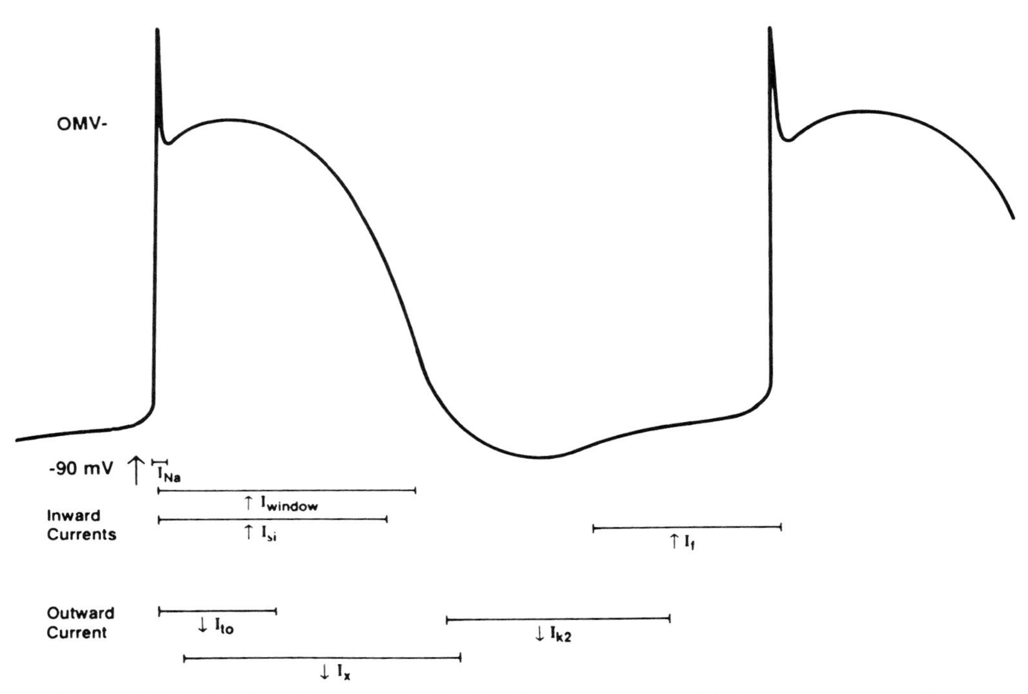

Figure 7.2 An idealized drawing of Purkinje fiber action potentials in a spontaneous firing preparation. At the bottom is a graphic representation of the approximate time course of the gated conductances. These are only rough approximations at best.

Phase 2

This phase is known as the *plateau phase*, in which there is a balance of currents. Slow inward currents are caused by calcium and sodium and outward currents are caused by chloride and potassium.

Phase 3

This phase is known as the *late repolarization phase*, and, during this period, the balance changes such that the inward currents decrease and the outward potassium current predominates. It becomes possible to stimulate the cell to depolarization during this phase.

Phase 4

When repolarization is complete, phase 4 begins. In atrial and ventricular myocardial cells there is no net movement of charge during phase 4; rather, there is a balance between slow inward sodium current and on outward potassium current. In contrast, Purkinje cells may demonstrate spontaneous diastolic depolarization. The membrane potential during phase 4 gradually decays to less negative values because of reduction in the outward potassium current, leaving only the inward or depolarizing sodium current. This spontaneous depolarization occurring during phase 4 is known as *automaticity* and, when it is seen in "slow responders," may be due to a slow inward calcium current rather than the loss of outward potassium current. The sinus nodal cells have this type of action potential and normally have the most rapid diastolic depolarization, resulting in the most rapid automatic rate in the heart.

Cardiac Currents

Another way of describing the action potential of cardiac cells is through a description of cardiac currents and their terminology. The following description of the most commonly discussed currents is intended as a summary and not as an exhaustive review.

I_{na} Current

This is known as the fast sodium current and is the large inward current that produces the rapid upstroke of the cardiac action potential. This upstroke is the driving force for cell-to-cell propagation of the action potential. Generally, the larger this current is, the greater the velocity of the upstroke and the more rapidly this waveform propagates through cardiac tissue. Activation and inactivation kinetics of this fast sodium current are highly voltage dependent.

I_{to} Current

The transient outward current was originally thought to be carried out by chloride ions, but in 1977 Kenyon and Gibbons (2) showed that potassium ions were the more likely candidate. I_{to} current is largely responsible for the notch seen in Purkinje fiber action potentials and occurs immediately after the upstroke in the earliest part of the plateau phase. This current is strongly frequency dependent.

I_{si} Current

The slow inward current (also known as the inward calcium current) is activated at potentials positive to -50 mV. This appears to be the primary inward current responsible for the action potential in central atrioventricular nodal cells, since calcium current blockers almost completely eliminate these action potentials (3). In normal atrial and ventricular myocardium, however, this current provides excitation by maintaining the action potential plateau and is important for contraction by providing the calcium trigger for activating the myofilaments.

I_k Current

This delayed outward plateau current (4) appears to be carried by potassium with contribution from other ions. It appears to have a major influence on action potential duration.

I_{ki} Current

This is the current that provides the resting potential in cardiac cells. It is a voltage-dependent outward current that is carried largely by potassium. A small inward leak current, primarily carried by sodium, slightly reduces the membrane potential from the equilibrium potential for potassium that is predicted by the Nernst equation.

Currents and Ion Channels

The currents described above occur via ion channels. Ion channels primarily control the electrical potential across the cell membrane, and therefore, it is important to understand not only the components of the cell membrane but also the formation of ion channels and their role in production of electrical current that results in the action potential. The remainder of this chapter deals with these issues.

Cell Membrane

The cell membrane, a lipid bilayer composed of phospholipid molecules, separates the internal aqueous environment of the cell from the extracellular space and limits the free movement of charged ions (such as calcium) between the extracellular and intracellular compartments. This restraint of ion passage allows intracellular ion concentrations to be different from those of extracellular fluid. The inner and outer aspects of the bilayer consist of hydrophilic ("water-loving") charged groups that permit interaction between the extracellular and intracellular fluid compartments. Centrally, the bilayer is composed of hydrocarbon chains of phospholipid fatty acids and is very hydrophobic ("water-hating"). This hydrocarbon prevents the movement of charged particles such as calcium. Because of the structure of the cell membrane, specialized channels that span the lipid bilayer are necessary for transport of such ions. The cell membrane is more than a simple insulator, in that it contains many different receptors that enable the cell to respond to humoral and transmitter stimuli. These receptors are primarily proteins that bind specific agonists. Cells communicate with their environment through these receptor proteins on the cell membrane.

There are two major types of ion channel. First, there are channels (i.e., Na^+ channels) that are opened or "gated" by simple membrane depolarization ("voltage dependent"). A second subset of channels require ligand binding to activate them (i.e., acetylcholine-K^+ channel) and open the "gate." Some ion channels are themselves receptors, whereas others are linked to other receptors through guanine nucleotide binding proteins, often referred to as "G-proteins." Ion channels can be considered efficient enzymes that allow certain types of ions to flow across the cell membrane. They control intracellular concentrations of ions, and these concentrations control cell functions such as muscle contraction and neurotransmitter secretion, and other processes including cell division. Translocation of ions through some of these channels does not require much energy, since they open and allow ions to flow down their concentration gradient. Many channels are highly selective and allow only a single type of ion to flow through the membrane. Like enzymes, these ion channels have regulatory sites that can be on either the extracellular or the intracellular side of the membrane or even within the pore of the channel. Since ion channels are relatively large (100 to 500 kD) compared to other membrane proteins, many possible regulatory sites exist.

Ion channels, as previously noted, primarily control the electrical potential across the membrane. The interior of each cell has a potential of -70 to -90 mV with respect to the outside. Inward cationic currents depolarize the cell to more positive potentials, whereas outward currents hyperpolarize the cell to more negative potentials. The resting membrane potential approaches the equilibrium potential for potassium because, in the steady state, more potassium channels are open than other kinds of channels. When a channel opens, it helps drive the membrane potential to its equilibrium potential: for potassium this is -90 mV, for sodium it is 55 mV, for calcium 150 mV, and for chloride -30 to -80 mV. The opening and closing of ion channels thus underlie all changes in membrane potential, including the well-known action potentials in cardiac membranes. In all cells, the changes in membrane potential appear to be important signaling functions. Therefore, understanding the regulation of ion channels is critical to understanding cell signaling.

The remainder of this chapter describes the various ion channels in the heart and their molecular properties. Additionally, because of the importance and relationship of ion channels to G-proteins, this topic is also covered.

Sodium Channels

Since the upstroke of the action potential (phase 0) in the heart results from influx of sodium ions through voltage-sensitive, sodium-selective pores (also known as sodium channels), these channels are of extreme importance to cardiac electrophysiology. Unlike sodium channels in other tissues, however, cardiac sodium channels also contribute to the action potential plateau phase (phase 2), as shown by the observation that submaximal blocking concentrations of tetrodotoxin could curtail the plateau without changing the maximal upstroke velocity (5,6). Hence, there is even greater importance in understanding these channels at the molecular level.

Sodium channels appear to belong to a separate and distinct subtype of channels, differing from calcium and potassium channels. Sodium conductance was first described by Hodgkin and Huxley (7) in neurons when they postulated that changes in sodium permeability during nerve action potentials could be described as the voltage-dependent opening and closing of "gates" for movement of sodium ions. The opening and closing reaction of these conductance gates, according to the Hodgkin and Huxley model, operates via two independent mechanisms, both of which respond directly to the membrane potential: 1) an activation gate that opens rapidly, and 2) an activation gate that closes slowly upon step depolarization. Membrane sodium conductance was shown to rise rapidly as a result of activation, reaching a peak within a few milliseconds before gradually decaying as the channel inactivates. The probability of occupancy in one of these states is time dependent and voltage dependent. As a result of the voltage dependence of sodium channel conductance, the current-voltage relationship shows a steep negative resistance region between −60 and −20 mV that is responsible for the all-or-none feature of the propagated action potential. Repolarization restores both activation and inactivation mechanisms to their resting configuration. The membrane conductance appears to be the sum of the random activity of single channels.

Kunze and co-workers (8) and Kirsch and Brown (9) have shown, using single-channel studies, that cardiac sodium channel inactivation proceeds with at least two rate constants. The fast phase of inactivation allows channels to open briefly (1 to 2 ms) without reopening. Less frequently, channel openings occur in bursts lasting several milliseconds and may account for the slow component seen in whole cell currents. Additional, even slower, components of inactivation appear to give rise to late sodium currents during the action potential plateau. Cardiac sodium channels appear to consist of a kinetically heterogeneous population with different channel subtypes. These subtypes appear to have different conductances, with the main conductance level being 20 PS (8) and other levels being of approximately 60% normal amplitude. Whether low-conductance events represent a substrate of the main conductance or openings of a separate set of channels is not yet known. It is possible, however, that these channel types are independent and therefore two separate channel types may exist. The origin of the late sodium current that contributes to the cardiac action potential plateau was first ascribed to channels that open in prolonged bursts with long mean open times within that burst (10). Bursting of this type was attributed to temporary loss of inactivation in channels with the same unitary conductance as normal channels.

A second type of gating modification is known to be controlled through interaction with β-adrenergic receptors. Measurements of the maximal upstroke velocity have shown that, in partially depolarized ventricular muscles, application of β-adrenergic agonists (such as isoproterenol) (11,12) or cyclic adenosine monophosphate (cAMP)–potentiating agents (such as phosphodiesterase inhibitors) (13) causes decreased sodium conductance. The underlying mechanism of this effect appears to be a holding, potential-dependent shift of steady state inactivation along the voltage axis by as much as -20 mV after application of isoproteronol (14). This modification of gating has been mimicked in cell-free membrane patches by application of GTPcS, a nonhydrolizable guanosine triphosphate (GTP) analogue, or the GTPcS-activated GTP binding protein, known to mediate β-adrenergic stimulation of adenyl cyclase when adenosine triphosphate (ATP) is not present. These agents, therefore, appear to be responsible for modulating cardiac sodium currents, and this modulation likely occurs through both a cAMP-dependent phosphorylation of the channel and a direct interaction of the GTP analogue with the channel. Adrenergic modulation of cardiac sodium channels may have important physiological consequences. By inhibiting sodium current, the activation of β-adrenergic receptors can have a depressant effect on conduction, in addition to its well-known stimulatory effect. This depressant effect would depend on both membrane depolarization and catecholamine concentration. During myocardial ischemia, for instance, both requirements for the depressant effect are fulfilled: 1) depolarization of the myocardium by increasing extracellular potassium concentration, and 2) increasing circulating catecholamine concentration. Depression of sodium current, slowed conduction, and reentrant arrhythmias could cause fatal ventricular fibrillation (15,16).

Moorman and co-workers (17) have provided evidence for a modulation of sodium currents through yet another mechanism, activation of angiotensin II receptors. This is consistent with a negative shift of the voltage dependence of activation along the voltage axis. The effect of angiotensin II is mimicked by phorbol ester–induced activation of protein kinase C (PKC) and is eliminated by preincubation in phorbol ester, a treatment that has been shown to down-regulate PKC (18). These results have suggested that modulation of cardiac sodium channel gating by activation of angiotensin II receptors proceeds via a second messenger pathway mediated by PKC. The effect of angiotensin II receptor on sodium currents would be expected to result in a decrease in action potential threshold and an increase in reentrant arrhythmias. This modulatory mechanism may have clinical relevance since it is known that patients with chronic congestive heart failure have decreased incidence of ventricular arrhythmias when treated with drugs that block conversion of angiotensin I to angiotensin II (19).

Molecular Biological Analysis of Sodium Channels

It has been shown in the brain that there are at least three separate genes that code for different calcium channels, and it is likely there is more than one cardiac or skeletal muscle calcium channel. The sodium channel was first isolated from the eel electroplax by Agnew and co-workers in 1978 (20). It is clear now that sodium channels have considerable homology but also express considerable tissue specificity. The sodium channels characterized for the brain are different from those of skeletal and cardiac muscle. Sodium channels, at that time, were found to be heavily glycosylated membrane-bound proteins consisting of a single 270-kD polypeptide. Similar results have been found in both rat and chick hearts, in which the sodium channel was shown to consist of a single 230- to 270-kD glycopeptide (21,22). By contrast, both rat brain and rat skeletal muscle sodium channels consist of large α subunits of 260 to 270 kD linked to one or more smaller β subunits of 35 to 45 kD (23,24). The role of these smaller subunits is not clear since functional channel reconstitution can be achieved with the α subunit alone (25). It is believed by many that β subunits may stabilize the structure (26) and modify the gating of the α subunit (27). The lack of β subunits in cardiac sodium channels may be a factor responsible for distinct gating behavior and pharmacological properties. Another factor may be the differences in posttranslational processing. These may be seen in rat brain and skeletal muscle sodium channels, which have high sialic acid content in comparison to rat heart sodium channels (21). Yet another factor in distinguishing sodium channel subtypes is the difference in primary amino acid sequences. Genes encoding sodium channels from rat brain and eel electroplax have been cloned, providing the amino acid sequence of α subunits (28). In rat brain, three different complementary DNAs (cDNAs) have been isolated whose sequences show 80% to 90% homology and 60% homology compared with that of eel (29,30). The fact that different sodium channel subtypes are present in rat heart and skeletal muscle is indicated by the observation that RNA probes developed from nonhomologous regions of these sodium channel cDNAs from the brain hybridize weakly or not at all with the mRNA from the heart and muscle, while much stronger hybridization is obtained using a probe corresponding to a homologous region.

The cDNA first isolated from the electric eel for the voltage-gated sodium channel was obtained using immunoscreening of a cDNA expression library. The deduced amino acid sequences from partial clones obtained from this library were compared with sequences from tryptic peptides; the deduced amino acid sequence did not show a signal peptide and the NH_2-terminus was found to be on the cytoplasmic side of the membrane. The predicted secondary structure and membrane topology were derived from hydropathy plots (31) and algorithms that predict secondary structure (32). The model that evolved shows four homologous domains, each containing six hydrophobic segments (S1 through S6) (29) (Fig. 7.3). The most conserved domain is the positive-charged segment S4, with an arg or lys at every third position. This segment is present in all voltage-gated sodium, calcium, and potassium channel cDNAs isolated thus far and has been proposed as the voltage sensor. The best functional evidence for this comes from a study in which arg or lys from the putative domain 1 was mutated to neutral or negatively charged amino acids individually or in combination. Mutants showed

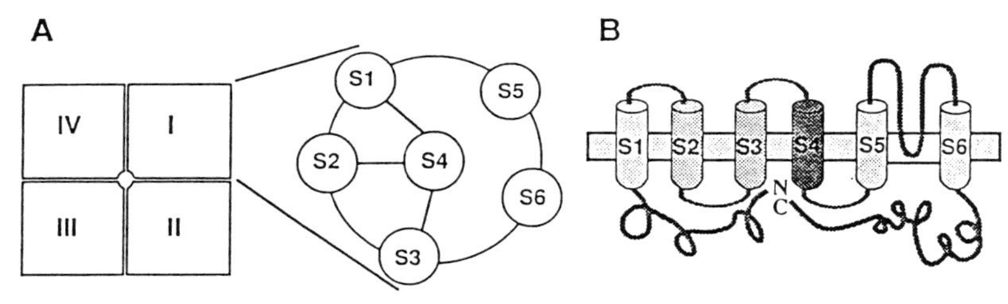

Figure 7.3 Idealization of Na^+ channel based on studies by Noda et al. (28, 29). (A) A two-dimensional representation laid out in a planar bilayer. The channel has four membrane repeats consisting of six segments each, all highly homologous. S4 has net positive changes of 4, 6, 7, and 8 in repeats I through IV consecutively; it is the proposed voltage sensor or gate. The NH_2- and COOH-termini are cytoplasmic (IN side). (B) A cross section made through the channel at the level of the cytoplasmic connecting loops. The *right-hand side* shows a proposed arrangement (28). The θ is S4 and S2, with its net negative charge, forms the hydrophilic wall of the channel.

significant change in conductance-voltage relationship of a type consistent with S4 being the activation site (33). Transcripts from the electric eel cDNA did not express in *Xenopus* oocytes, whereas those from sodium channel cDNAs from rat brain did express currents in oocytes. Three distinct cDNA clones were isolated from rat brain using eel cDNA probes and had overall structures similar to the eel channel. Rogart and coworkers (34) cloned a cDNA encoding a cardiac sodium channel in 1989 and subsequently expressed it in *Xenopus* oocytes. The current was expressed at low levels and the concentration of tetrodotoxin required to block it, a hallmark for differentiating cardiac from neuronal channels, was submicromolar. A glu at residue 386 was shown to correspond to a glu at residue 387 in brain channels, whereas glu 387, when mutagenized to a neutral amino acid, demonstrated tetrodotoxin insensitivity. Mutagenesis of asn 388 in brain has a corresponding arg at 387 in the heart channel. This could repel the positively charged guanidinium group of tetrodotoxin and confer lower sensitivity on the cardiac sodium channel. Finally, the α subunits of voltage-gated sodium and potassium channels have been found to be quite similar. The amino acid sequence has been interpreted by models of tertiary structure as having four similar domains, with each domain having six to eight transmembrane segments, all of which are probably α-helices. Gene duplication probably gave rise to the four domains, since there is more homology between different domains of the two proteins than there is among different domains of either protein.

Workers in the area of cardiac sodium channels are presently attempting to understand the relationships of pharmacological agents to these channels. Cardiac sodium channels appear to be 50 to 1,000 times more sensitive than neuronal channels to blockage by local anesthetics, such as lidocaine. These anesthetics are effective antiarrhythmic agents that selectively block impulse activity from damaged or depolarized tissue without affecting normal rhythm generation and conduction. They do so by binding with high affinity to open or depolarization-inactivated channels and slowing their recovery from inactivation, thereby preventing impulse generation from the damaged region. Resting noninactivated channels are much less susceptible to blocks since they quickly recover from the inactivated state. Although not proven at this time, local anesthetics appear to act at specific sites on the channel environment. Indirect evidence argues for a specific site since local anesthetics both block sodium current and noncompetitively inhibit batrachotoxin binding, with both effects being saturable and stereospecific (35,36). This supports the specific receptor concept. Recent efforts to identify binding sites of cardiac sodium channels by certain toxins has succeeded in isolating two different 18–amino acid segments of the α subunits (37). These sites, therefore, mark extracellular regions of the channel that can interact with inactivation gating. Whether the toxin binding site on the cardiac sodium channel has a similar location and amino acid sequence homology remains to be determined. Further work in this area is ongoing.

Calcium Channels

All muscle cells are dependent upon rapid changes in the concentration of intracellular calcium ions for contraction to occur. The concentration of free calcium in muscle cells increases from the resting state during excitation and contraction. Different types, however, have different dependency and responsiveness to calcium entry from extracellular fluid. Skeletal muscle contraction depends largely on intracellular calcium storage and release, whereas contraction of cardiac or smooth muscle cells is initiated by entry of calcium through the sarcolemma. Influx of calcium into muscle cells occurs through specific protein channels in the sarcolemma. Voltage changes across the sarcolemma trigger the passage of calcium through these channels and, therefore, these channels are considered voltage dependent or voltage sensitive. Since calcium is a charged ion, its influx into the muscle cell is associated with an electrical current, and this current plays an important role in cardiac function. Calcium ions entering through cardiac calcium channels are essential for excitation-contraction coupling since they initiate the normal cardiac contraction by triggering calcium release from the sarcoplasmic reticulum. Under pathological conditions, calcium channels can generate arrhythmic activity by initiating slow action potentials in parts of the heart in which such electrical activity is not normally found. The inward current carried through open calcium channels helps determine action potential duration and refractory period in all cardiac cells, diastolic depolarization and cardiac rhythm in pacemaker cells, and the action potential upstroke and conduction velocity in nodal cells. Calcium currents in cardiac tissue were first reported in 1973 by Reuter (38), and this information was expanded upon through information gained with the introduction of the patch clamp technique by Neher and Sakmann (later improved by Hamil et al. in 1981 [39]), which allows recording of calcium channel activity at the level of individual channel molecules. Neher and Sakmann received the Nobel Prize for this work in 1991.

The calcium channel is a complex glycoprotein that traverses the lipid bilayer of the sarcolemma. These glycoproteins determine the physiological properties of the channel, including selectivity for a particular ion species and gating, which allows opening and closing of the channel to determine ion passage at only specific times and physiological conditions. The cardiac sarcolemma of most animal species contains two calcium channel types that share high selectivity for calcium over other ions but differ in all other important properties (40–44). The two channel types are most commonly referred to as L-type and T-type calcium channels (42), and differ in voltage range of their activation and inactivation as well as in their microscopic (i.e., single channel) and macroscopic (i.e., whole cell) gating kinetics.

The "L" of the L-type channels stands for large conductance and long-lasting current. The "T" of the T-type calcium channels stands for tiny conductance and transient current. The differences in the voltage dependence of activation and inactivation are useful for distinguishing these two channel types in whole cell recordings (41). In contrast to the combination of slow, voltage-dependent and fast, calcium-dependent mechanisms of inactivation of L-type channels, which explains why L-type currents inactivate more rapidly when carried by calcium, the inactivation for T-type channels depends only on voltage. Single-channel currents carried by L-type channels are larger than those carried by T-type channels. In addition, the two channel types differ in pharmacological sensitivities. L-type current is increased by organic calcium channel agonists and decreased by antagonists such as nitrendipine. β-adrenergic stimulation is also known to increase L-type current. T-type channels are blocked by amiloride, whereas L-type channels are unaffected. In addition, T-type channels are insensitive to dihydropyridine calcium channel agonists and antagonists and fail to respond to β-adrenergic agonists (41,42,45,46).

The L-type calcium channel is the classically described calcium channel (38,47). Because of its kinetics it is assumed to be the main calcium channel contributing maintained inward current to the action potential plateau. Because of their relatively positive range of inactivation, L-type channels are likely to determine the rate of rise to slow action potentials occurring in the depolarized region of the atrioventricular node, a conclusion supported by the fact that atrioventricular conduction is very sensitive to organic calcium channel antagonists acting specifically on L-type channels. L-type channels alone are able to deliver the calcium required for normal excitation-contraction coupling, as evidenced by the fact that normal intracellular transients and contractions can be observed from holding potentials at which T-type calcium channels should be completely inactivated (48). Furthermore, in some animals (i.e., rabbit, frog) L-type channels are the only calcium channels found in ventricular cells.

The physiological role of T-type calcium channels in cardiac tissues is less well understood. Because of their negative range of inactivation, these channel might help set firing thresholds in myocardial and Purkinje cells. A possible involvement of T-type channels in the generation of pacemaker potential has also been hypothesized (41,42). The striking functional differences between the two calcium channel types make it likely that they are separate molecular entities or that they share common protein cores to which different regulatory subunits are linked. In 1987 Cooper and co-workers (49) reported the subunit composition of L-type channels. The relative density of the two channel types differs between atrial and ventricular cells (41), with L-type channels being larger in ventricular cells while T-type channels are comparable to the L-type channels in atria. This difference could be explained if L-type channels are preferentially located in the membranes of T-tubules while T-type channels are present throughout the surface membrane. Such a preferential localization has been shown for L-type channels in skeletal muscle (50), but no evidence in cardiac muscle has thus far been obtained.

The calcium channels face special challenges in having to selectively transport calcium, an ion vastly outnumbered by other extracellular ions, and to transport it at high enough rates to ensure proper delivery of the calcium-encoded message. As judged by reversal potentials under defined ionic conditions, both L- and T-type channels have permeability ratios for divalent-monovalent ions on the order of greater than 10^3:1 (1,000:1) (51). Such high selectivity has been explained by the presence of multiple intrapore high-affinity binding sites for calcium (52). A high-affinity intrapore binding site for calcium may explain why, at millimolar extracellular calcium concentrations, the pore is always occupied by calcium and thus effectively impenetrable by other ions. The high affinity of calcium is also consistent with tight binding and, therefore, limited mobility for calcium.

Calcium Channel Molecular Biology

Calcium channels are functionally more diverse and structurally more complex than sodium channels. The skeletal muscle T-tubule calcium channel has been studied the most extensively and is a tetrameric structure consisting of α_1, α_2, β, and γ components (53). A γ subunit is disulfide bonded to α_2 and is encoded by the α_2 gene (54). Dihydropyridines (DHPs) have been used as calcium blockers and were used to purify the DHP receptor (DHPR) from skeletal muscle. This purified DHPR from skeletal muscle has been incorporated into phospholipid bilayers, and active DHP-sensitive calcium channels were recovered that retained physiological and pharmacological properties of a functional L-type channel. The α_1 subunit has 1,873 amino acids (molecular mass: 212 kD) and its hydrophobicity profile is similar to that of sodium channels, giving a predicted general organization similar to sodium channels. The cDNA for the α_2 subunit encodes 1,106 amino acids (molecular mass: 125 kD) with a 26–amino acid signal sequence at the NH_2-terminal end, which is extracellularly located. There are 18 possible *N*-glycosylation sites and two cAMP-dependent phosphorylation sites. The α_2 subunit has no known homology to known ionic channels or receptor proteins. α_1 and α_2 subunits are expressed differentially in various tissues. Northern analysis using RNA from various tissues revealed that α_1 transcripts were present in skeletal muscle, although weak signals came from the aorta and heart as well. The same Northern blots using RNA from the ileum and brain revealed no signal for α_1 transcripts, but the α_2 transcript was detected in RNA from all tissues studied. In 1989, Mikami and co-workers (55) cloned the α_1 DHPR from heart tissue and expressed convincing calcium currents in *Xenopus* oocytes. These currents were increased when α_2 transcripts were co-injected. The α_2 subunit of skeletal T-tubules is heavily glycosylated. It has been cloned, and its characterization demonstrates hydrophilic sequences without the usual structures seen in voltage-dependent channel proteins (56).

Ruth and co-workers recently cloned the β subunit of skeletal muscle DHPR and deduced its primary sequence (57). This sequence had no homology with any known proteins and was consistent with that of a peripheral membrane protein. This subunit is phosphorylated, not glycosylated, and is thought to be cytoplasmic in location. The γ-subunit, in contract, was recently cloned (58) and shown to be a glycoprotein having four putative transmembrane domains and two *N*-linked glycosylation sites. The transcript is not evident in hybridizations utilizing heart or brain RNA.

Even less is known about the subunit composition of cardiac DHPR. Cardiac α_1 mRNA expresses calcium currents in oocytes, but it is uncertain if other subunits are present in these oocytes. DHPs have been widely used to label calcium channels because they are present in picomolar quantities in skeletal muscle and because their photoactivated derivatives (which bind covalently) have been synthesized. These derivatives have been used so extensively that the DHPR has become synonymous with the calcium channel. The best studied calcium channel is the DHPR of rabbit skeletal T-tubules. The molecular masses of the α_1, α_2, β, γ, and δ subunits are 170, 150, 52, 32, and 25 kD, respectively, and, under nonreducing conditions, the α_2 and δ subunits appear to be disulfide bonded, creating a 175-kD subunit. Binding sites for the DHPs, verapamil and diltiazem are located on the α_1 subunit and all are allosterically linked. The α_1 subunit is phosphorylated by protein kinase A (PKA), calcium calmodulin-dependent kinase (59), cyclic guanosine monophosphate (cGMP)–dependent protein kinase, PKC (60), and a protein kinase intrinsic to the skeletal muscle triads (61). The β subunit is also phosphorylated by cAMP-dependent kinase and PKC. The phosphorylation rates are different, with PKA working faster on α_1 subunits and PKC more rapid on β subunits. In α_1 subunits, ser 687 is a consensus site for PKA phosphorylation and is the site most rapidly phosphorylated.

The α_1 subunit cDNA of rabbit skeletal muscle encodes a protein that is approximately 40 kD larger than the α_1 subunit protein purified from muscle (approximately 190 kD). It appears that some posttranslational processing occurs, and in skeletal muscle the differences may account for the proteins destined to serve as voltage sensors and those destined to serve as calcium channels (62). The cardiac mRNA is 2 to 3 kilobase pairs (kb) larger than the skeletal muscle transcript based on Northern analysis, with the greatest difference being an extramembrane region at the NH_2 and COOH-termini (55,63). The cardiac α_1 subunit protein is also larger than its skeletal counterpart. Alignment of the primary sequence of the two deduced proteins has an overall homology of 66%, with homology of the putative membrane segments being higher. As a result, the phosphorylation sites at which modulation occurs are likely to be different.

Calcium currents in skeletal muscle are influenced by neurotransmitters despite the fact that the receptors are on the surface sarcolemma, whereas calcium channels are mainly on the T-tubules. The abundance of consensus sites for phosphorylation by PKA in the α_1 subunit predicts cAMP-dependent effects on single calcium currents. Marked increases in the opening probability occur without changes in the unitary current amplitudes or mean open times (63–65). In the heart, where β-adrenergic simulation is more important, maximal stimulation by β-agonists produces fivefold increases in calcium currents. Trautwein and co-workers (64) have worked out the pathway for β-adrenergic stimulation. At high concentrations, approximately 80% to 90% of the effect is mediated by the pathway of G_s (the stimulator of adenylyl cyclase) adenylyl cyclase itself, cAMP, PKA, and the calcium channel. Inhibitors of cAMP phosphodiesterase such as isobutylmethylxanthine, which have effects on cGMP phosphodiesterase, also increase calcium currents (66). However, increase in cGMP may cause decreased calcium currents in certain species (such as in frog heart) by increasing the activity of cAMP phosphodiesterase. This is not the case, however, for the mammalian heart since cGMP causes decreased sodium currents, which were increased by nonhydrolyzable 8-bromo-cAMP (67). These cGMP effects are most effective after increasing calcium currents with β-agonists. At lower concentrations, a direct stimulatory effect of G_s is prominent (68). A direct, membrane-delimited pathway from G_S to calcium channels has been identified in cardiac tissue. In cardiac tissue compared with skeletal muscle, it has been proposed that G_S is more potent for calcium channels than for adenylyl cyclase. It is believed that a direct pathway is involved in beat-to-beat regulation of heart rate, since buildup of cAMP via a cytoplasmic pathway (that is found only in skeletal muscle) is likely to be slow (Fig. 7.4).

Muscarinic agonists such as acetylcholine decrease cardiac calcium currents after these currents have been increased by β-agonists (69). These muscarinic agonists have little effect on the basal calcium current, however. The effect appears to be at the level of adenylyl cyclase, since stimulation by cAMP is not affected whereas stimulation by forskolin, which acts directly on adenylyl cyclase, is blocked. The mediator is a G-protein and the effect is mediated by both α subunits and β-γ dimers.

The concensus sequences of PKC in the cardiac α_1 subunit predict that PKC activators such as phorbol esters will have effects on calcium

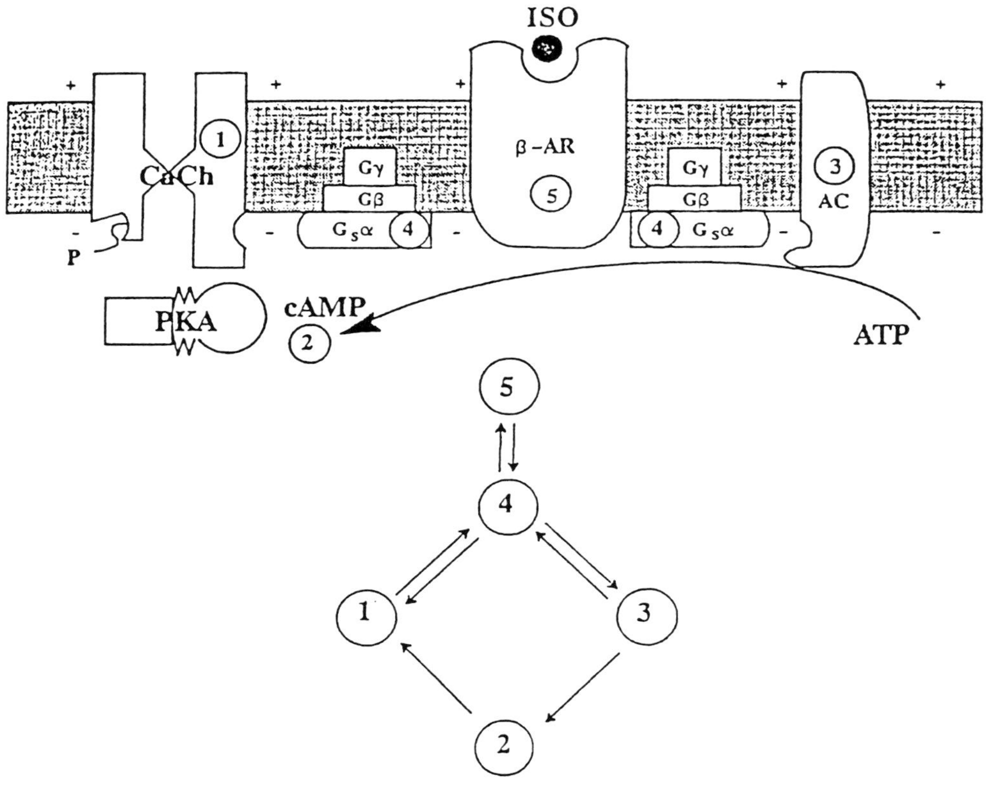

Figure 7.4 Pathway from Gs to calcium channels. β-*AR*, β-adrenergic receptor; *G*, heterotrimeric guanine nucleotide binding proteins; *AC*, adenylyl cyclase; *ATP*, adenosine triphosphate; *CaCh*, high-threshold Ca^{2+} channel; *PKA*, cAMP-dependent protein kinase.

currents that are biphasic (70). Initially, stimulation resulting from an increased opening probability occurs; this is later followed by a decrease. The initial stimulation is voltage dependent, whereas the subsequent decrease occurs as a result of down-regulation and is specific for phorbol esters. Prior exposure to these PKC activators will abolish the transient increase in opening probability. The PKC effect may be the mechanism by which angiotensin changes cardiac calcium currents (71), and it has been proposed that α-adrenergic agonists may affect currents via this mechanism as well (72).

Thus, the basic understanding of calcium channels has clearly increased dramatically over the past decade. The new knowledge gained via molecular biological methods has increased the understanding of the role played by calcium channels and many of the effects caused by pharmacological agents working at the level of the calcium channel. The future appears to hold the key for increasing the ability of physicians and scientists to better control the contractility and electrical currents occurring via these calcium channel mechanisms. This understanding will continue to improve our ability to manage patients with diseases of the heart caused by abnormal calcium currents and lead to pharmacological therapy of cardiac disease.

Potassium Channels

Potassium is the major ion determining the resting membrane potential, which depends on the intracellular potassium concentration being high (150 mmol/liter; nearly 40 times higher than the extracellular concentration of this ion) and on the potassium channel itself. Maintenance of this high intracellular concentration of potassium must occur in order to maintain the resting membrane potential. Compared with sodium and calcium channels, understanding of the molecular basis of potassium channels and their requirements has been gained slowly. Although there is a paucity of data, the following paragraphs will outline what is presently known about potassium channels.

Voltage-gated potassium channel cDNAs encode a single domain, but the arrangement is similar to each domain of the sodium and calcium channels. For this reason, a tetrameric structure for the potassium channels is presently considered most likely. However, the diversity of potassium channels appears to be greater than that seen with either sodium or calcium channels. The initial cDNA cloning of potassium channels was performed using the skeletal muscle of a mutant *Drosophila*, called the Shaker (73). This cDNA was obtained from a large gene with numerous exons, and alternative splicing of this gene was shown to produce five distinctly different mRNAs with a common core structure; each RNA expressed a different phenotype (74,75). Probes from these potassium channels have been used to screen a cardiac tissue library, which resulted in several clones being obtained that express currents similar to those expressed by neuronal and muscle potassium channel cRNAs (76,77). These currents resemble the typical cardiac potassium current but are clearly different from the cardiac delayed rectifier potassium channel current. A second type of potassium channel cDNA has also been cloned from renal epithelial cells (78). Study of the cDNA sequence demonstrates that the open reading frame encodes a small protein of approximately 10 kD with a single transmembrane segment. Expression of this clone produces a slow-activating potassium current with kinetics similar to those of a delayed rectifier potassium channel in the heart. A similar cDNA was most recently cloned from heart muscle, and its cRNA produces currents having slow kinetics of cardiac delayed rectifier potassium channels (77).

It is likely that more genes for cardiac ion channels will be found since there are many different phenotypes. Included in these are the inwardly rectifying, ligand-gated potassium channels, which are gated through sodium and ATP and have thus far eluded scientists.

An extremely important discovery was made when direct coupling between G-protein and ion channels was first demonstrated. It initially seemed most likely for a muscarinic M2 receptor and the potassium-acetylcholine (KAch) channel to be coupled. Supported by the findings that 1) latency occurs after acetylcholine application (a latency that appears to be much longer than the latency found at the nicotinic acetylcholine receptor, where the receptor and ion channel were the same protein), and 2) such second messengers as cGMP, cAMP, and calcium were not involved, this direct coupling appeared increasingly plausible. When acetylcholine was applied outside the cell-attached patch of membrane, it could not activate potassium channels, whereas acetylcholine in patch pipettes could. Binding studies showing that muscarinic M2 receptors were linked to G-proteins in the heart further supported this hypothesis (79). The link with the electrophysiology of cardiac cells was established when blockage of muscarinic effect on inwardly rectifying whole cell current and resting membrane potential by PTX was shown, along with the requirement for intracellular GTP (80). Additional support was provided when it was found that a nonhydrolyzable GTP analogue disconnected the currents from the acetylcholine control (81). Another set of experiments was performed and demonstrated that a GTP analogue applied to the cytoplasmic face of an inside-out membrane patch activated a specific set of inwardly rectifying, single potassium channel currents, and that these were the same as those activated by acetylcholine. It was also shown that it did so in a magnesium-dependent manner (82). Yatani and co-workers (83) subsequently demonstrated that a G-protein, when preactivated with this GTP analogue, activated single-channel KAch currents in precisely the same manner as physiological activation. Phosphorylation by PKC was also excluded. Little was known about the site of protein-protein interaction at which the α subunit activates potassium channels. The best information was provided by trypsin activation of muscarinic potassium channels, which produced single-channel currents in which inward rectification, single-channel conductance, mean open time, and burst duration were indistinguishable from muscarinic activation. Since trypsin is known to inactivate G-proteins, it was thought that its effect is probably on the potassium channel or structures closely associated with it. Trypsin cleaves proteins at lys or arg residues, whereas the arg-specific reagents glyoxal and phenylglyoxal do not activate potassium channels. Based on these data, the hypothesis that trypsin disrupted an inhibitory gating mechanism that normally holds the channel closed in the absence of activated G-protein was advanced. The inhibitory gates were believed to be physically distinct from the gate that mediated bursting and were shown to contain at least one trypsin cleavage point located at a lys residue accessible from the cytoplasmic surface of the cell membrane.

Potassium-ATP (KATP) current has recently been shown to be regulated by G-proteins, but the pathways that couple receptor, G-protein, and channel have not been defined. A variety of studies have led to the belief that KATP channels may be coupled to adenosine receptors via these proteins.

G-Proteins

G-proteins are made up of three protein subunits of approximately 40, 35, and 11 kD, and these are known as α, β, and γ, respectively. The α subunit contains a single binding site for GTP and mediates the GTPase activity that accompanies signal transduction by these proteins. There are several subtypes of α, β, and γ subunits, giving rise to different types of G proteins each with distinct functional roles. The functional differences in G-proteins arise primarily out of the differences found in the α subunit; for example, G-proteins containing an α_5 (G_5) subunit stimulate adenylyl cyclase, and those containing an α_1 (G_1) subunit inhibit adenylyl cyclase. The brain exhibits a high concentration of G-protein that neither stimulates nor inhibits adenylyl cyclase (called G_0), which contains a third type of α subunit known as α_0. Transducin (G_t), a G-protein found in the retina, is activated by rhodopsin and mediates transduction of light signals. G_t again contains an α_1 subunit; in addition, however, it contains a γ subunit different from that in G_1, G_0, or G_5. At least 10 distinct α subunits, two β subunits, and three γ subunits have been identified, all occurring in various combinations and giving rise to numerous types of G-proteins. There is, at this time, good evidence that a common G-protein mediates the ability of adenosine and muscarinic receptors to activate the current known as I_{KAch}. This current occurs as a result of stimulation of the parasympathetic nerves to the heart, which then releases a quantum of acetylcholine to generate inhibitory postsynaptic potentials mediated by a potassium current. The action of acetylcholine on this current is dependent on the presence of internal GTP (84). A GTP analogue will activate I_{KAch} even without the presence of an agonist (85). Therefore, there appears to be close coupling between I_{KAch} channels and a regulatory G-protein (86). Cardiac cells appear to have at least four types of G-proteins, including G_5, G_1, and two others that react with antibodies for G_0 (87). The density of G_1-like and G_0-like G-proteins in the adult dog heart is four to five times that seen in muscarinic receptors (88). Additionally, a 20- to 30-fold excess of G_1-like and G_0-like proteins over those of muscarinic receptors has been reported in embryonic chick hearts (89). Cardiac membranes appear to contain many more receptors than channels, and more G-proteins than receptors. A functional unit, such as that observed in an excised membrane patch, may thus consist of one or more channels, 200 or more receptors, and 1,000 or more pertussis-sensitive G-proteins, including G_1-like and G_0-like proteins.

Interest in the molecular biology of potassium channels and their interaction with G-proteins has recently become a topic of significant interest. With the linkage data provided by Keating and co-workers (90) on Romano-Ward long QT syndrome (LQTS), which demonstrated linkage in a large Mormon family of LQTS to the short arm of chromosome 11 (11p) in the region of 11p15.5, where the Harvey-*ras*-1 (H-*ras*-1) oncogene is localized, the possibility of a major disease being caused by abnormalities of potassium channel–G-protein interaction has come into play. The H-*ras*-1 oncogene acts as a G-protein with significant interaction with potassium channels, specifically the KAch channel. This linkage has recently been confirmed in 7 of 11 families studied by Towbin and co-workers (J.A. Towbin et al., personal communication, 1991). The cloning of the gene responsible for the long QT syndrome and the characterization of the protein product are presently underway. Thus, in the not-too-distant future, it appears that specific abnormality in this system might be shown to be responsible for causation of an important human cardiac disorder responsible for a large number of sudden deaths in children and young adults. Unlocking the keys to disorders such as LQTS is likely to allow for improved clinical understanding and care for patients with hereditary and acquired conductor abnormalities.

References

1. Ling G, Gerard RW: The normal membrane potential of frog sartorious fibers. *J Cell Comp Physiol* 1949;34:383.
2. Kenyon JL, Gibbons WR: Four-amino pyridine and the early outward current of sheet cardiac Purkinje fibers. *J Gen Physiol* 1979;73:139.
3. Zipes DP, Mendez C: Action of manganese ions and tetrodotoxin on atrioventricular nodal transmembrane potentials in isolated rabbit hearts. *Circ Res* 1973;32:447.
4. Noble D, Tsien RW: Outward membrane currents activated in the plateau range of potentials in cardiac Purkinje fibers. *J Physiol* 1969;200:205.
5. Attwell D, Cohen I, Eisner D, et al: The steady-state TTX-sensitive ("window") sodium current in cardiac Purkinje fibres. *Pflugers Arch* 1979;379:137.
6. Coraboeuf E, Deroabaix E, Coulombe A: Effect of tetrodotoxin on action potentials of the conduction system in dog heart. *Am J Physiol* 1979;236:H561.
7. Hodgkin AL, Huxley AF: A quantitative description of membrane current and its application to conduction and excitation in nerve. *J Physiol (Lond)* 1952;117:500.
8. Kunze DL, Lacerda AE, Wilson DL, Brown AM: Cardiac Na currents and the inactivating, reopening and waiting properties of single cardiac Na channels. *J Gen Physiol* 1985;86:691.
9. Kirsch GE, Brown AM: Kinetic properties of single sodium channels in rat heart and rat brain. *J Gen Physiol* 1989;93:85.

10. Patlack JB, Ortiz M: Slow current through single sodium channels of adult rat heart. *J Gen Physiol* 1986;86:89.
11. Arita M, Kiysue T, Aornine M, Imanishi S: Nature of "residual fast channel" dependent action potentials and slow conduction in guinea pig ventricular muscle and its modification by isoproterenol. *Am J Cardiol* 1983;52:1433.
12. Hisatome I, Kiyosue T, Imanishi S, Arita M: Isoproterenol inhibits residual fast channel via stimultaion of β-adrenoceptors in guinea-pig ventricular muscle. *J Moll Cell Cardiol* 1985;17:657.
13. Windisch H, Tritthart HA: Isoproterenol, norepinephrine and phosphodiesterase inhibitors are blockers of the depressed fast Na^+-system in ventricular muscle fibers. *J Mol Cell Cardiol* 1982;14:431.
14. Schubert B, Van Dungen AMJ, Kirsch GE, Brown AM: Inhibition of cardiac Na^+ currents by isoproterenol. *Am J Physiol* 1990;258:14977.
15. Goldstein DS: Plasma norepinephrine as an indication of sympathetic neural activity in clinical cardiology. *Am J Cardiol* 1981;48:1147.
16. Hirche HJ, Franz CH, Bos L, et al: Myocardial extracellular K^+ and H^+ increase and nonadrenaline release as possible cause of early arrhythmias following acute coronary artery occlusion in pigs. *J Mol Cell Cardiol* 1980;12:579.
17. Moorman JR, Hirsch GE, Lacerda AD, Brown AM: Angiotensin II modulates cardiac Na^+ channels in neonatal rat. *Circ Res* 1989;65:1804.
18. Shenolikar S, Karbon EW, Enna SJ: Phorbol esters down-regulate protein kinase C in rat brain cerebral cortex slices. *Biophys Biochem Res Commun* 1986;139:251.
19. Webster MWI, Fitzpatrick A, Nicholls MG, et al: Effect of enalapril on ventricular arrhythmias in congestive heart failure. *Am J Cardiol* 1985;56:566.
20. Agnew WS, Levinson SR, Brabson JS, Raftery MA: Purification of the tetrodoxin-binding component associated with the voltage-sensitive sodium channel from *Electrophorus electricus* electroplax membranes. *Proc Natl Acad Sci USA* 1978;75:2606.
21. Gordon D, Merrick D, Wollner DA, Catterall WA: Biochemical properties of sodium channels in a wide range of excitable tissues studied with site-directed antibodies. *Biochemistry* 1988;27:7032.
22. Lombet A, Lazdunski M: Characterization, solubilization, affinity labelling and purification of the cardiac Na^+ channel using Tityus toxin γ. *Eur J Biochem* 1984;151:651.
23. Meissner DJ, Catteral WA: The sodium channel from rat brain separation and characterization of subunits. *J Biol Chem* 1985;261:10597.
24. Roberts R, Barchi RL: The voltage-sensitive sodium channel from rabbit skeletal muscle: Chemical characterization of subunits. *J Biol Chem* 1987;262:2298.
25. Noda M, Ikeda T, Suzuki H, et al: Expression of functional sodium channels from cloned cDNA. *Nature* 1986;322:826.
26. Meissner DJ, Catteral WA: The sodium channel from rat brain: Role of the β_1 and β_2 subunits in saxitoxin binding. J Biol Chem 1986;261:211.
27. Auld VJ, Goldin AL, Rafte DS, et al: A rat brain Na^+ channel and subunits with novel gating properties. *Neuron* 1988;1:449.
28. Noda M, Shimizu S, Tanabe T, et al: Primary sequence of *Electrophorus*

electricus sodium channel deduced from cDNA sequence. *Nature* 1984; 312:121.
29. Noda M, Ikeda T, Kayano T, et al: Existence of distinct sodium channel messenger RNAs in rat brain. *Nature* 1986;320:188.
30. Kayano T, Noda M, Flockerz V, et al: Primary structure of rat brain sodium channel III deduced from the cDNA sequence. *FEBS Lett* 1988;228:187.
31. Kyte J, Doolittle RF: A simple method for displaying the hydropathic character of a protein. *J Mol Biol* 1982;157:105.
32. Garnier JD, Osquothorpe DJ, Robson B: Analysis of the accuracy and implications of simple methods for predicting the secondary structure of globular proteins. *J Mol Biol* 1978;120:97.
33. Stuhmer W, Conti F, Suzuki H, et al: Structural parts involved in activation and inactivation of the sodium channel. *Nature* 1989;339:597.
34. Rogart RB, Cribbs LL, Muglia LK, et al: Existence of distinct sodium channel messenger RNAs in rat brain. *Nature* 1989;86:8170.
35. Sheldon RS, Cannon NJ, Duff HG: A receptor for type I antiarrhythmic drugs associated with rat cardiac sodium channels. *Circ Res* 1987;61:492.
36. Sheldon RS, Cannon NJ, Nies AS, Duff NJ: Sterospecific interaction of tocainide with the cardiac sodium channel. *Mol Pharmacol* 1988;33:327.
37. Tejedor FJ, Catteral WA: Site of covalent attachment of α-scorpion toxin derivatives in domain I of the sodium channel α-subunit. *Proc Natl Acad Sci USA* 1988;85:8742.
38. Reuter H: Divalent ions as charge carriers in excitable membranes. *Prog Biophys Mol Biol* 1973;26:1.
39. Hamil O, Marty A, Neher E, et al: Improved patch-clamp techniques from high-resolution current recording from cells and cell-free membrane patches. *Pflugers Arch* 1981;391:85.
40. Bean BP: Classes of calcium channels in vertebrate cells. *Annu Rev Physiol* 1989;51:367.
41. Bean BP: Two kinds of calcium channels in canine atrial cells: Differences in kinetics, selectivity and pharmacology. *J Gen Physiol* 1985;86:1.
42. Nilius B, Hess P, Lansman JB, et al: A novel type of cardiac calcium channel in ventricular cells. *Nature* 1985;316:443.
43. Mitra R, Morad M: Two types of calcium channels in guinea pig ventricular myocytes. *Proc Natl Acad Sci USA* 1986;93:5340.
44. Hagiwara N, Irisawa H, Kameyama M: Contribution of two types of calcium currents to the pacemaker potentials of rabbit sino-atrial node cells. *J Physiol (Lond)* 1988;395:233.
45. Bonvallet R: A low threshold calcium current recorded at physiological Ca concentration in single frog atrial cells. *Pflugers Arch* 1987;408:540.
46. Tytgat J, Nilius B, Vereeke J, et al: The T-type Ca channel in guinea-pig ventricular myocytes is insensitive to isoproterenol. *Pflugers Arch* 1988; 411:704.
47. Tsien RW: Calcium channels in excitable cell membranes. *Annu Rev Physiol* 1983;45:341.
48. Wier WG, Isenberg G: Intracellular [Ca^{2+}] transients in voltage clamped cardiac Purkinje fibers. *Pflugers Arch* 1982;392:284.
49. Cooper CL, Vandaele S, Barhanin J, et al: Purification and characterization

of the dihydropyridine-sensitive voltage-dependent calcium channel from cardiac tissue. *J Biol Chem* 1987;262:509.
50. Fosset M, Maimbrich E, Delpont E, et al: [$_3$H]Nitrendipine receptors in skeletal muscle: Properties and preferential localization in transverse tubules. *J Biol Chem* 1983;258:6086.
51. Lee KS, Tsien RW: High selectivity of calcium channels in single dialysed heart cells of the guinea pig. *J Physiol (Lond)* 1984;354:253.
52. Almers W, McCleskey EW: Non-selective conductance in calcium channels in frog muscle: Calcium selectivity in a single-file pore. *J Physiol* 1984;353:585.
53. Campbell KP, Leung AT, Sharp AH: The biochemistry and molecular biology of the dihydorpyridine-sensitive calcium channel. *Trends Neuro Sci* 1988;11:425.
54. DeJongh KS, Warner C, Catterall WA: Subunits of purified calcium channels: α_2 and β are encoded by the same gene. *J Biol Chem* 1990;265:14738.
55. Mikami A, Imoto K, Tanabe T, et al: Primary structure and functional expression of the cardiac dihydropyridine-sensitive calcium channel. *Nature* 1989;340:230.
56. Ellis SB, Silliams ME, Says NR, et al: Sequence and expression of mRNAs encoding at α1 and α2 subunits of a DHP-sensitive calcium channel. *Science* 1988;241:1661.
57. Ruth P, Roehrkaster A, Biel M, et al: Primary structure of the β-subunit of the DHP-sensitive calcium channel from skeletal muscle. *Science* 1989;245:1115.
58. Jay SD, Ellis SB, McCue AF, et al: Primary structure of the γ-subunit of the DHP-sensitive calcium channel from skeletal muscle. *Science* 1990;248:490.
59. Hosey MM, Lazdunski M: Calcium channels: Molecular pharmacology structure and regulation. *J Membrane Biol* 1988;194:81.
60. Jahn H, Nastaincyzk W, Roehrkasten A, et al: Site-specific phosphorylation of the purified receptor for the calcium-channel blockers by cAMP- and cGMP-dependent protein kinase protein kinase C, calmodulin-dependent protein kinase II and casein kinase II. *Eur J Biochem* 1988;178:535.
61. Ingawa T, Leung AT, Campbell KP: Phosphorylation of the 1,4 dihydropyridine receptor of the voltage-dependent Ca^{2+} channel by an intrinsic protein kinase in isolated triads from rabbit skeletal muscle. *J Biol Chem* 1987;262:8333.
62. Catterall W: Structure and function of voltage-sensitive ion channels. *Science* 1988;242:50.
63. Pelzer D, Cavalie A, Trautwein W: Cardiac Ca channel currents at the level of single cells and single channels. *Basic Res Cardiol* 1985;80:65.
64. Trautwein W, Cavalie A, Flockerzi V, et al: Modulation of calcium channel function by phosphorylation in guinea pig ventricular cells and phospholipid bilayer membrane. *Circ Res* 1987;61:117.
65. Renter H, Cachelin AB, DePeyer JE, Kokubun S: Modulation of calcium channels in cultured cardiac cells by isoproterenol and 8-bromo-cAMP. *Cold Spring Harbor Symp Quant Biol* 1983;XLVIII:193.
66. Levi RC, Alloatt G, Fischmeister R: Cyclic GMP regulates the Ca channel current in guinea pig ventricular myocytes. *Pflugers Arch* 1989;413:685.

67. Thakkar J, Tony SG, Sperelakis N, Wahler GM: Inhibition of cardiac slow action potentials by 8-bromo-cyclic GMP occurs independently of changes in cAMP levels. *Can J Physiol Pharmacol* 1988;66:1092.
68. Yatan A, Brown AM: Rapid β-adrenergic modulation of cardiac calcium channel currents by a fast G protein pathway. *Science* 1989;245:71.
69. Hescheler J, Kameyana M, Trautwein W: On the mechanism of muscarinic inhibition of the cardiac Ca current. *Pflugers Arch* 1986;407:182.
70. Lacerda AE, Rampe D, Brown AM: Effects of protein kinase C activators on cardiac Ca^{2+} channels. *Nature* 1988;335:249.
71. Dosenici A, Dhallan RS, Cohen NM, et al: Phorbol ester increases calcium current and stimulates the effects of angiotensin II on cultured neonatal rat heart myocytes. *Circ Res* 1988;62:347.
72. Hartmann HA, Magossa NJ, Kleiman RB, Houser SR: Effects of phenylephrine on calcium current and contractility of feline ventricular myocytes. *Am J Physiol* 1988;255:H1173.
73. Tempel BL, Papajian DM, Schwartz TL, Jan YN, Jan LY: Sequence of a probable potassium channel component encoded at a Shaker locus of *Drosophilia*. *Science* 1987;237:770.
74. Jan LY, Jan YN: Voltage-sensitive ion channels. *Cell* 1989;56:13.
75. Kamb A, Iverson LE, Tanovye MA: Molecular characterization of Shaker, a *Drosophilia* gene that encodes a potassium channel. *Cell* 1987;50:405.
76. Pfaff SL, Tamkun MM, Taylor WL: A *Xenopus* oocyte proteins expression vector. *Anal Biochem* 1990;188:192.
77. Folander K, Smith JS, Stein RB, Swanson R: Cloning of K^+ channels underlying the cardiac I_k current. *J Mol Cell Cardiol* 1990;22:(suppl 1):S1.
78. Murai T, Kukizuka A, Takumi T, et al: Molecular cloning and sequence analysis of human genomic DNA encoding of novel membrane protein which exhibits a slowly activating potassium channel activity. *Biochem Biophys Res Commun* 1989;161:176.
79. Rosenberger LB, Roeske WR, Yamamma HI: The regulation of muscarinic cholinergic receptors by guanine nucleotides in cardiac tissue. *Eur J Pharmacol* 1979;56:179.
80. Pfaffinger PJ, Martin JM, Hunger DD, et al: GTP-binding proteins couple cardaic muscarinic receptors to a K channel. *Nature* 1985;317:536.
81. Breitwieser GE, Szaba G: Uncoupling of cardiac muscarinic and β-adrenergic receptors from ion channels by a guanine nucleotide analogue. *Nature* 1985;317:538.
82. Kurachi Y, Nakajima T, Sugimoto T: On the mechanism of activation of muscarinic K^+ channels by adenosine in isolated atrial cells: Involvement of GTP-binding proteins. *Pflugers Arch* 1986;407:264.
83. Yatani A, Codina J, Brown AM, Birnbaumer L: Direct activation of mammalian atrial muscarinic potassium channels by GTP regulatory protein Gk. *Science* 1987;235:207.
84. Ribalet B, Ciani S, Eddlestone GT: ATP mediates both activation and inhibition of K(ATP) channel activity via cAMP-dependent protein kinase in insulin-secreting cell lines. *J Gen Physiol* 1989;94:693.
85. Parent L, Coronado R: Reconstruction of the ATP-sensitive potassium channel of skeletal muscle: Activation by a G-protein-dependent process. *J Gen Physiol* 1989;94:445.

86. Kurachi Y, Nakajima T, Sugimoto T: Acetylcholine activation of K^+ channels in cell-free membrane of atrial cells. *Am J Physiol* 1986;252:H681.
87. Scherer NM, Toro MJ, Entman MI, et al: G-protein distribution in canine cardiac sarcoplasmic reticulum and sarcolemma: Comparison to rabbit skeletal muscle membrane and to brain and erythrocyte G-proteins. *Arch Biochem Biophys* 1987;259:431.
88. Vatner DE, Lee DC, Schwarz KR, et al: Impaired cardiac muscarinic receptor function in dogs with heart failure. *J Clin Invest* 1988;81:1836.
89. Subers EM, Liles SC, Luetje CW, et al: Biochemical and immunological studies on the regulation of cardiac and neuronal muscarinic acetylcholine receptor number and function. *TIPS* 1988;9:25.
90. Keating M, Atkinson D, Dunn C, et al: Linkage of a cardiac arrhythmia, the long QT syndrome, and the Harvey *ras*-1 gene. *Science* 1991;252:704.

Index

Note: Page numbers followed by f refer to illustrations, page numbers followed by t refer to tables.

D

E

F

G

N

O

P

R

S